COMMON SENSE LEADERSHIP MATTERS

TOXIC LEADERSHIP DESTROYS
A Case Study

Security Disclaimer:

This book was submitted to the Department of Defense Office of Prepublication and Security Review on September 11, 2022. It was cleared for release to the public on March 17, 2023.

CLEARED AS AMENDED
March 17, 2023:
Department of Defense
OFFICE OF PREPUBLICATION AND SECURITY REVIEW

In an effort to protect specific organizations, operations, and/or the individuals who took part in the events covered in this book, there are a few instances where I had to use pseudonyms for people and generic names for military organizations such as the one I was a part. To comply with Department of Defense redactions I had to adjust some descriptions of pictures and events. The security-instituted adjustments to this book were minor and did not affect the way the stories were written or the lessons that were learned from them.

COMMON SENSE LEADERSHIP MATTERS

TOXIC LEADERSHIP DESTROYS
A Case Study

(Book III)

PETE BLABER

Book Design by Alan Barnett
Copy Edited by Pam Susemiehl
Maps by Mike McCarthy

www.peteblaber.com

ISBN: 979-8-218-07656-6 (paperback)
ISBN: 979-8-218-22483-7 (hard cover)

TABLE OF CONTENTS

PROLOGUE
What's This Book About?   vii

INTRODUCTION
Pat Tillman Saved My Life Last Night   1

PART I: HOW COMMON SENSE LEADERS CREATE A HEALTHY LEADERSHIP CLIMATE

CHAPTER 1
Walking in the Woods Under
A Healthy Leadership Climate   11

CHAPTER 2
What a Healthy Leadership Climate Looks Like   35

PART II: WHAT HAPPENED TO PAT TILLMAN AND HIS PLATOON IN AFGHANISTAN

CHAPTER 3
How Part II is Organized   45

CHAPTER 4
The Intersection of 2/75 in Iraq   51

CHAPTER 5
What is Toxic Leadership?   75

CHAPTER 6
How a Toxic Leadership Climate
Leads the Platoon Into Chaos   87

CHAPTER 7
Five Facts About Emotional Stress
and How to Counter It   153

CHAPTER 8
The Firefight   163

CHAPTER 9
The Aftermath and Discipline   253

CONCLUSION
Thoughts and Recommendations   273

ANNEX A
Military Unit Size and Leadership Rank   281

ANNEX B
Ranger Weapons Information   282

Notes   285

PROLOGUE

This is the third in a series of three leadership books. The argument of this book is contained in its title: *Common Sense Leadership Matters, Toxic Leadership Destroys*. In support, it provides a real-world case study[1] of what happened to Pat Tillman and his Ranger platoon in Afghanistan. As you'll learn from the pages that follow, it's not what any of us were told. Before getting to the story and the never-before-heard leadership lessons we can learn from it, it's important to first understand who Pat Tillman was.

Pat Tillman was born on 6 November 1976, in San Jose, California. The eldest of three boys, Pat was a caring son, a protective brother, and a natural leader with a tendency to push limits—in life, in the classroom, and in sports.

He attended Arizona State University on a football scholarship, where he led his team to the 1997 Rose Bowl after an undefeated season and three consecutive selections to the Pac-10 All-Academic Football Team. Described by many as a voracious reader with an unquenchable appetite for knowledge, Pat earned a B.S. in marketing, graduating Summa Cum Laude in three and a half years.

Although many considered him too small to play professional football, the Arizona Cardinals selected Pat in the seventh round of the 1998 NFL Draft. He answered the skepticism by becoming the team's starting safety and broke the franchise record for single-season tackles in 2000 with 224.

The day after the attacks of September 11, 2001, Pat told a reporter, "At times like this, you stop and think about just how good we have it, what kind of system we live in, and the freedoms we are allowed. A lot of my family has gone and fought in wars, and I really haven't done a damn thing."

In May 2002, Pat walked away from his $3.6 million NFL contract with the Cardinals and enlisted in the United States Army a few weeks later. The decision shocked many and garnered national media attention despite his refusal to speak publicly about the choice. Pat was a man of principle, and he stuck by his principles of "not making it about him" and "not treating his military service any differently than any other soldier."

Pat wasn't the only Tillman who was moved by the 9/11 attacks; his brother Kevin, a minor-league baseball player, decided to join the army with him. The Tillman brothers were assigned to the 2nd Battalion, 75th Ranger Regiment, in Fort Lewis, Washington, and did combat tours in Iraq in 2003, followed by Afghanistan in 2004.[2]

On the evening of 22 April 2004, Pat Tillman's Ranger platoon was ordered to split into two sections by their chain of command. Pat was with the first section, and his brother was traveling with the second section. Minutes after they were split apart, the second section of the platoon was ambushed as it traveled through a slot canyon in the rugged, cross-compartmented mountains of eastern Afghanistan. Pat Tillman's section of the platoon, which had passed through the canyon minutes earlier, left their vehicles and moved on foot to higher ground in an effort to provide covering fire for their fellow Rangers. An intense firefight followed, during which Pat Tillman and an Afghan soldier were shot and killed, and two other Rangers were seriously wounded.

Initially, the Army reported that Corporal Pat Tillman had been killed by enemy fire. Controversy ensued five weeks later, on 28 May 2004, when the Pentagon notified the Tillman family that he had actually been killed by friendly fire. The family and other critics alleged that the Department of Defense delayed the disclosure until weeks after Tillman's memorial service out of a desire to protect the image of the U.S. military, and the careers of the senior officers involved.

INTRODUCTION

The purpose of the pages that follow is not to judge, adjudicate, or assign blame. The purpose is to learn from what happened in the past so we can adapt to future contingencies and prevent similar situations like this one from ever happening again. If we don't learn, we can't adapt. If we can't adapt as individuals, organizations, and as a species, then we perish.

A great addition to the life legacy of Pat Tillman would be that the lessons learned from this incident would be shared and passed on for many generations to come. And that ten, twenty, and even fifty years from now, future leaders will continue to build on and add to his legacy when they apply what they learned to their own life-or-death survival situations and proclaim: *"Pat Tillman saved my life last night."* A lofty aspiration for sure, yet one that makes the sacrifices and effort needed to accomplish it even more worthwhile.

On 25 April 2017, I received an email that stated:

A mutual friend told me that you have information about Pat's death. I would like to speak to you if you're willing.

It was signed by Mary Tillman (Pat's mother). Later that day, I called and talked with Mary on the phone. I explained that our mutual friend from the Rangers told me that she was still gathering information and trying to find answers about what had happened to her son. And even though I wasn't sure the information I had would be helpful, I felt that I had a duty to share it with her in case it turned out to be relevant to other information she was learning. I then told her that I had been in Iraq at the same time (April-May 2003) and place (Baghdad International Airport) as her sons and their unit.

During that time frame, three senior sergeants from her sons' unit—the 2nd Ranger Battalion—who I had known and respected for many years, approached me to ask for advice and assistance regarding the battalion's leadership climate. Although I was no longer in the 2nd Ranger Battalion, they approached me out of desperation because, as one of them described it to me: "the leadership climate is so toxic, and everyone is so stressed out about it, that I'm concerned something really bad is going to happen."

Mary appreciated the information, even though it had happened in Iraq a full year before her son was killed in Afghanistan. The main point she emphasized to me was that despite four investigations, hundreds of investigative articles, and a couple of books, she still didn't know what actually happened to her son during the firefight that killed him. Specifically, she didn't know whether Pat was killed by accident or on purpose. She had been lied to so many times by the people involved that she didn't know who or what to believe.

"This isn't about me," she continued in a calm, composed tone, "it's about every parent that has lost a child and doesn't understand why. I can't be the only one. I don't want another mother to go through what I've gone through. If the situation was reversed and something had happened to me, Pat would never rest until he learned the truth about what happened." Then she asked me: "If you have time and think it worthwhile, could you take a look at the investigation documents and tell me what you think?"

I had lots of questions and was dawn-to-dusk busy with my job, my family, and writing my second book (*The Common Sense Way*, 2021). However, when the mother of a fellow soldier who was killed in action asks you for assistance in sorting out the details of what happened to her son, life's day-to-day trivialities somehow seem a lot less important. No matter how busy or demanding my schedule, I'm not sure I'm capable of giving any other answer than, "I'd be happy to help" and "when can I see the documents."

The next day I drove four hundred miles to meet with Mary at her workplace, and take possession of a large, plastic storage container packed to the brim with declassified documents from all four Army investigations into the death of Pat Tillman. When I arrived at her office, she walked outside to meet me in the parking lot during her lunch break. We shook hands, and I wasn't sure what to say, so I told her how I felt: "Mrs. Tillman, first and foremost, I want to tell you how sorry I am about what happened to your son, and as a former Army officer and former Ranger, I also want to apologize for the way the military treated you and your family. Every member of the military I've ever talked to about this is ashamed and embarrassed about what they did and how they treated you."

She thanked me and then told me, "Thirteen years after my son was killed, I still don't know what really happened out there. I read your book, and I know you were a Ranger and that you were in Afghanistan. I trust your opinion, so please let me know what you think happened and how we can keep it from happening again." To which I responded, "Mrs. Tillman, I have no idea what actually happened to your son, but I will read every document in this container and I'll talk to as many former Rangers as I can connect with, and when I'm finished, you have my word that I'll tell you exactly what I think happened, even if it's not what you may want to hear." She graciously replied, "That's why I contacted you.

I want to know what you think happened, not what you think I want to hear."

We talked for about an hour, and she had to get back to work, so I loaded the storage container in the back of my SUV, said good-bye to Mary, and drove four hundred miles back to my house. Early the next morning, I opened the storage container, picked out a random binder, and opened it up to read a few pages. In addition to the investigation documents, the container also included schematics of the vehicle configurations and locations, as well as maps, photos, and videotapes of the terrain in and around the area where the firefight took place. Nine hours later, I was still reading, highlighting, and attaching the first of what would turn into hundreds of sticky notes.

As I continued to read and cross-reference the investigation documents, I also began talking to the individual Rangers of 2nd Platoon. My first big takeaway was what a complex and dangerous situation the Rangers of 2nd Platoon had been in that day. The firefight took place in the mountains of eastern Afghanistan, near the border with Pakistan, in an area that is best known for its difficult and daunting terrain, hostile tribes, and xenophobic people.

My second big takeaway was that there were a number of key decisions made during the hours before the platoon was ambushed that seemed to defy common sense. These decisions (e.g. splitting the platoon, and dragging an inoperable Humvee through some of the most difficult and dangerous terrain on the planet), played a key role in the chaos, confusion, and friendly fire that followed.

My third big takeaway emerged while talking with the individual Rangers about those fateful key decisions. I was shocked at how little they knew about what had actually happened (most were not given the opportunity to see the results of the four official investigations), as well as the long-term devastation this incident had on so many of them. It was an eye-opening experience for me with regards to how severe the PTSD problem can be within certain groups of military combat veterans (e.g., same unit, same place, same timeframe, same battle, etc.). Although the firefight and follow-on investigations ended many years ago, the platoon was still taking casualties from both. The Rangers' inability to understand and explain what had actually happened and why, left them with feelings of frustration and guilt, the twin pillars of PTSD.

I never allowed myself to take a position on who was at fault or what specific action might have caused any part of the incident, and I think that helped me in the discovery process. **The question that drove my research was, "What really happened, and what can we, the people, learn from it?"**

When I asked my friend Ted Kennedy, a retired 2nd Ranger Battalion sergeant major, for his thoughts and/or recommendations on what I should do with the information I was learning, here's what he said: "Remember when we were just starting out back in the late 80s and early 90s? Guys like us read everything we could get our hands on about lessons learned from past combat operations. Books like, This Kind of War, Once an Eagle, Dispatches, We Were Soldiers Once and Young, and Hackworth's book[3]. Remember what kind of influence those books and their lessons had on us? Today, the guys like us are our kids, and they are the ones that are searching for knowledge and real-world examples that they can learn from and maybe even use when they find themselves in similar situations in the future. We have an ongoing responsibility to future leaders to ensure they learn from situations like this; otherwise, more good men will die in the future like they already had before this incident happened."

Curious, I asked him to clarify what he meant by "like they already had." He told me about an incident that happened during the 2nd Ranger Battalion's previous deployment to Afghanistan in November of 2003, which led to the death of another highly respected Ranger named Jay Blessing. "Most of the same leader-ship lessons from the Tillman incident should have already been learned from the Blessing incident, but no one ever talked about and shared those lessons, so it turned out we didn't learn a thing." Additional details surrounding the "Jay Blessing incident" are dis-cussed in Chapter 6.

My friend's wise words regarding our "on-going responsibility to future leaders" made me realize I had a duty to write about what I discovered and to share what we, the people, can learn from it. My first inclination was to make it the final chapter of my sec-ond book, *The Common Sense Way*, which was half-finished when I received the Tillman documents. However, as the weeks and months of writing and research accumulated, the "final chapter" page count went from 50 to 100, to 150, to 250+, before including maps and photos. Somewhere along the way, I realized I was writ-ing a third book.

Over the ensuing months, I talked with Mary Tillman a few more times on the phone to ask her questions and to give her updates on what I was learning. It wasn't always what she wanted to hear, but she always listened. Sometimes she laughed, some-times she sighed, but she never tried to influence my thoughts or my takeaways from what I learned.

I recontacted Mary Tillman to tell her I was about to submit this book to the Department of Defense Office of Pre-Publication Review and to ask if she'd like to do a pre-read. She politely declined and told me she didn't want to do a pre-read because she didn't want to influence what I found or what I wrote. She then told me the same thing she did when we first connected: "This is why I contacted you. I want to know what you think happened, not what you think I want to hear."

My answer to Mary Tillman's initial question and request is contained on the pages that follow: A detailed chronology, reconstruction, and case study of what happened based on the four separate U.S. Army investigations, over one hundred interviews with members of the 2nd Ranger Battalion, as well as maps, illustrations, and photos of the area around the firefight.

The case study consists of two sections. The first, "How Common Sense Leaders Create a Healthy Leadership Climate" (30 pages), tells the story of how I learned what common sense leadership is while serving as a Lieutenant in the 2nd Ranger Battalion, as well as how to use common sense leadership to create a healthy and thriving leadership climate.[4]

The second section, "What Really Happened to Pat Tillman and His Platoon in Afghanistan" (235 pages), also takes place in the 2nd Ranger Battalion and contrasts with the first section by providing a cautionary example of what can happen to an organization and its people when common sense leadership is absent and toxic leadership fills the void.

Together, the two sections provide a real-world case study that reinforces why common sense leadership matters and how toxic leaders and the toxic leadership climates they create can destroy not only an organization's ability to successfully accomplish its purpose, but the organization itself, and in extreme situations, the people that are forced to live and work under them.

PART I

HOW COMMON SENSE LEADERS CREATE A HEALTHY LEADERSHIP CLIMATE

CHAPTER 1

Walking in the Woods Under a Healthy Leadership Climate[5]

Fort Lewis, Washington, South Rainier Training Area, 3:00 am November, Late 1980s: After five hours, there was no longer any doubt about it; we were lost. It was supposed to be a five-kilometer (three-mile) tactical movement. Over the past five hours, the platoon likely logged three times that distance. When it comes to humping heavy loads across thickly forested and hilly terrain, every step of extra distance is doubly frustrating.

Although the temperature was in the mid-40s, the wet Washington air and the ice-cold winds born over the glacial-fed waters of Puget Sound made the mid-40s feel like freezing. As long as you keep moving at a steady pace, you can work up enough body heat to fend off the freeze. You have to keep moving.

The last update from the platoon leader had passed back through the column three hours earlier: "The patrol base is a hundred meters up ahead." Five hundred meters later, the platoon leader had pretty much lost face. One thousand meters later (by my pace-count), the men began moving like molasses at a speed appropriately called painfully slow.

All forests are difficult to move through at night. Old growth forests like South Rainier distinguish themselves due to three unique features: the dark, the damp, and the deadfall. All three are progenies of the same family tree.

Giant conifers such as Sitka spruce, western red cedar, and Douglas fir dominate the forests of western Washington. Covered by moss and rising to towering heights of 250 feet and higher, the treetops create a cathedral-like canopy, blocking almost all

ambient light. Even when wearing night vision goggles, it's hard to see your hand in front of your face. Which makes seeing your map almost impossible.

Rain falls frequently in South Rainier, with total precipitation averaging between four and six feet each year. The result is a lush green canopy of shrubs, large leafy flora, plants called epiphytes that grow on top of other plants, and nurse logs, which are downed or dead trees that support new plant life. Every inch of the forest floor is either covered or coated in green. (See photo.) "Jungle" is the most common word people use when they first experience South Rainier, which explains why the U.S. Army used it in the 1960s and 70s to train and prepare new recruits for deployment to Vietnam.[6]

The final feature that makes movement in these forests both difficult and dangerous is deadfall. Deadfall is a tangled mass of fallen trees and branches that are randomly strewn across the forest floor and usually camouflaged by green undergrowth. Deadfall not only slows you down, it also takes a brutal toll on the shins. The bruises and abrasions got so bad at times that some Rangers wore soccer shin pads so they could avoid any new injuries and allow the old ones to heal. As an added bonus, the bruises and abrasions are almost always followed by a fall.

As a result, it was standard operating procedure to treat deadfall the same way as other hazards such as water obstacles, steep drop-offs, or poison plants. When the first person in the formation encountered deadfall on the trail, it was expected that they'd pass a detailed description back through the column as a warning for those that followed. Reflecting both the seriousness of the hazards and the genuine concern for the well-being of their fellow Rangers, the descriptions were often quite detailed, and in retrospect quite humorous: "Deadfall, twelve inches off the ground at a ninety-degree angle to the trail, multiple branches with sharp edges."

Helping the guy behind you helps him avoid getting hurt and helps the entire platoon move quicker and quieter. It also feels good.

If you're moving tactically and there's not enough light to see your map, the best way to keep track of where you are is by using your watch, your pace count, and your compass. Over the last two hours, the platoon had walked a total of 1,500 meters, during which we made a series of four ninety-degree left turns, each about 500 meters apart. Four ninety-degree turns in the same direction and same distance equals a square. You didn't need a math degree to understand what was happening. We were walking in circles. Still no word or explanation from the platoon leader.

"What is he doing up there? He has to know he's lost. Why the 'F' hasn't he updated the rest of the platoon?" I mumbled to myself. Cold, wet, tired, frustrated, and pissed off. The timeless Ranger recipe for a drone-fest.[7] Onward we trudged.

As the company executive officer (XO), I wasn't actually a member of the 2nd Platoon. The XO position is usually filled by the senior first lieutenant in the company. The XO's primary job is to assist the company commander (a captain) in mission planning and execution. In peacetime, the company XO is more accurately described as the company operations officer, responsible for coordinating all training and logistics such as food, ammo, vehicles, and aircraft, etc.

The company XO is technically second-in-command of the company, so for contingency purposes on any type of tactical movement, the company commander and the XO always split up (e.g., ride in separate vehicles, fly in separate aircraft, or walk with separate platoons, etc.). On this night, I was moving with 2nd Platoon while the company commander was moving with 1st Platoon. I guess you could call it "tactically tagging along."

The mission of the 2nd Platoon was to conduct a tactical movement to a location specified by the company commander, where they were to set up a patrol base[8] and bed down for the night in preparation for a new mission that they would receive in the morning. This was the first time I had the opportunity to move tactically with the 2nd Platoon. The platoon leader was relatively new, so I didn't have any firsthand knowledge about his leadership experience.

A platoon leader's job is to command the platoon, but it's the platoon sergeant and the squad leaders who control it. The platoon sergeant of the 2nd Platoon was a combat veteran and one of the most senior members of the battalion. The squad leaders were all top-notch. And the rest of the platoon, like the rest of the company, was stacked with some of the finest soldiers and sergeants in the Army. Which made the senselessness of our current situation even harder to fathom.

Clearly, they were having problems with their navigation. So why didn't I do something about it? During the first three hours of the movement, **I was trying to do unto them as I'd want any observer or higher-ranking person to do unto me. Stay out of their decision-making loop and let them operate the same as they would on a real-world mission. If they made mistakes or encountered problems, they should be allowed to work together as a team to learn how to fix them. That's how we learn to lead.** Having said that, there is no way I would have waited this long and allowed the platoon to get this lost if not for the fact that there was a significant barrier between the platoon leader and me.

I was moving in the back of the forty-man column while the platoon leader and platoon sergeant were both moving near the front. Using the military standard of five-meter intervals between men, multiplied by forty men, meant that the platoon column—which contracts and expands like an accordion as it moves through, around, and over restrictive terrain—was between 200 and 400 meters long. Getting from the back of the column to the front of the column while moving in darkness and densely forested terrain not only takes a lot of time but also creates a lot of commotion and confusion, which combine to make it tactically unsound.

I had a radio on my back, but making contact with another radio in this type of terrain required both radios to have their twenty-foot-long whip antennas attached and extended. Thanks to the jungle-like undergrowth, the only way to get the twenty-foot-long antenna up and extended was to do so while stationary. We had only stopped one time in the last five hours, during which I managed to get the antenna attached and extended but was unable to make contact with the platoon leader before the column started moving again. I'm not sure what I would have said to him or how I could have helped him on the radio anyway. What I really needed to do was to talk to him face-to-face with map in hand.

Should I stay or should I go? Instead of doing something about it, I chose to do nothing and suck up the pain.

Then it started to rain.

Followed a few seconds later by a human wave of hand and arm movements passed back through the column to signal a halt. "About F-ing time." It felt good to finally take a knee. Like thunder following lightning a few seconds later, I heard the faint rumble of "the word" as it was passed from Ranger to Ranger, from the front of the column to the rear, from the man in front to the man behind.

The Ranger in front of me was my link to the rest of the platoon and the only other person I had talked to over the last five hours. When he approached, I heard a strange sound. I figured out what it was as soon as he started to talk. "Sir, ch ch ch ch, word just got passed back from the platoon leader that we're almost there, so he wants everyone to don their Gore-Tex rain gear underneath their uniforms, ch ch, ch, ch."

It took a few seconds for my brain to cut through the chatter and comprehend what he just said. "Did you say 'underneath our uniforms,' corporal?" I replied as my teeth began to chatter too. "Yes, Sir, it's a platoon S.O.P. The platoon leader says that wearing Gore-Tex over our uniforms makes too much noise as it rubs together while we walk. He says wearing it underneath our uniforms helps muffle the noise and prevents the enemy from hearing us before we hear them."

Un-freaking believable, was what I wanted to say, but I kept my thoughts to myself as the corporal hurried back to his rucksack to start stripping down and bundling up.

As much as I wanted to charge forward through the column and challenge the platoon leader on the senselessness of his order, my chattering teeth reminded me that I had another priority. *There's no way I'm stripping down to my skivvies in the freezing rain to put my Gore-Tex on underneath my uniform. It doesn't make sense. I'll just wear my Gore-Tex top and bottoms over my uniform like they're meant to be worn.* I quickly pulled my Gore-Tex jacket out of my rucksack and had one arm in the sleeve when I thought about the corporal and the other Rangers in front of me. *If they have to wear their Gore-Tex underneath their uniform, how can I, with good conscience, walk behind them with mine on the outside?*

Then it started to pour.

Yet instead of putting my Gore-Tex jacket on as quickly as possible, I just stood there frozen in thought. Caught between making sense and senselessness, my mind was impaled on the horns of an ethical dilemma. The sound of the corporal's voice and a lot of commotion up ahead snapped me out of it. I flipped down my night vision goggles to see what I could see just in time to witness what appeared to be some sort of ancient tribal dance ritual. I was droning I knew. Of course, it wasn't a dance ritual. It was the four guys in front of me trying to take their pants off without taking their boots off first. It didn't work well when we were little kids, and it doesn't get any easier as your feet and boots get bigger.

After dropping their pants, each Ranger began hopping around on one foot while pulling and prying to get the non-hopping pant leg free from the boot. This was Exhibit A for why putting your Gore-Tex on under your uniform doesn't make tactical sense. Then one Ranger tipped over. Like bowling pins, when he tipped, he took out the Ranger next to him, who also tipped over and took out the other two. This was the first time I sensed the seriousness of our situation.

Then something unexpected happened. They started laughing. Hysterically. And I laughed with them. During miserable movements like this one, laughing at falls isn't considered rude or impolite. The wipeouts and the laughs that follow provide some of the only highlights from otherwise frustrating and forgettable long nights in the bush. It didn't last long. Once those that fell realized how much wetter and colder they were, the laughter faded, and anger and frustration returned to prominence.

The mud-groveling fall was both Exhibit B and a timely lesson for me. After giving a hand to the four guys groveling on the ground, I no longer had any doubt about what the right thing was to do, so I told them, "You guys need to put your Gore-Tex on over your uniforms as per company and battalion S.O.P." Then I added, "This is a safety issue, and if anyone asks you can tell them the XO told you to do it."

Unfortunately, it turned out to be a lesson only half-learned. In my rush to put my Gore-Tex on as quickly as possible, I boneheadedly attempted to step into my Gore-Tex bottoms without taking my boots off first. Balancing on one leg, I threaded my left boot through the left Gore-Tex pant leg and started to pull. These were the earliest versions of Gore-Tex rain suits, so they didn't come equipped with zippers or snaps that allowed you to loosen the bottom of the pant leg first. The only way to get them over my boot was to pull harder. Which I did. Then I tipped over and fell face-first onto the spongy moss that covered the forest floor. Then something not so unexpected happened. Howls of laughter erupted from my four fellow fallers. And once again, I laughed too.

Lying on the forest floor, sopping wet and freezing cold, with one boot stuck inside my Gore-Tex pant leg, made me realize how senseless my choices were leading up to that moment. "Get your shit together," I scolded myself. My timing was impeccable.

"Has anyone seen the XO?" someone whispered loudly. His tone was angry urgent. "XO, are you back here?" I scrambled to pull myself up and wipe the mud off my face, but it was too late. "XO? Is that you?" I took a deep breath and stood up, "Yes, it's me," I replied with all the self-deprecation I could muster. It was Sergeant X⁹, a squad leader at the time, probably twenty-four years old and already known as one of the finest and fittest warriors in the battalion. Although he never worked directly for me, he and I had conducted a couple of long-range reconnaissance missions together, and he was one of the NCOs I had had the good fortune of working with repeatedly throughout my time in the 2nd Ranger Battalion.

"Please don't tell me you were putting your Gore-Tex on under your clothes too?" I didn't answer. "I'd make fun of you, Sir, but I tried to do the same thing until I heard a strange noise in the woods behind me. I thought it was a critter sneaking up on me, and then I realized it was one of my guy's chattering teeth. He was so wet-cold he could barely talk, and that's when I realized this was starting to turn into a serious safety issue.

"I tried calling the platoon leader and platoon sergeant on the radio but wasn't able to make contact, so I figured I'd walk back here and talk to you about it. While making my way through the platoon column, I discovered the entire platoon is dangerously cold, tired, frustrated, and red-hot pissed off. Which is why everybody—including you, Sir—are starting to do some really stupid shit."

Sergeant X's feedback hit me like a punch in the face. It also ended my senseless debate. I shook my head as reality sunk in; *I waited way too long. Yes, I'm just tactically tagging along with the platoon tonight, but that doesn't mean I'm no longer a leader. As second-in-command of the company, these guys are my guys too.* All the reasons I had come up with to stay in the back of the column now seemed like a self-centered rationalization to do nothing: too hard to put up my long whip antenna, too much time, and too difficult to make my way through the column. What I should have been asking myself was, what would I want a leader in my position to do if I was one of the Rangers walking circles in the freezing rain? *Thanks for slapping me out of it, Sergeant X,* I thought to myself.

"Do you know where we are?" I asked Sergeant X with a hint of embarrassment. "I do now, Sir," he replied. "I foolishly stopped following along after the first hour because I couldn't see my map without turning on a light. I also figured there was no way we could possibly get lost on a movement this short. After an hour of zigzagging and no word on what was going on, I realized I needed to figure out where we were. Luckily, I was watching my compass and keeping a pace count, so I had one of my guys throw a poncho over my head while I used a finger light to figure it out. How about you, Sir?"

"I lost track of my pace count during the last ninety-degree turn. I was trying to subtract the distance we walked after the previous turn, but I was so pissed off I couldn't do the math. Can you show me where you think we are on the map?"

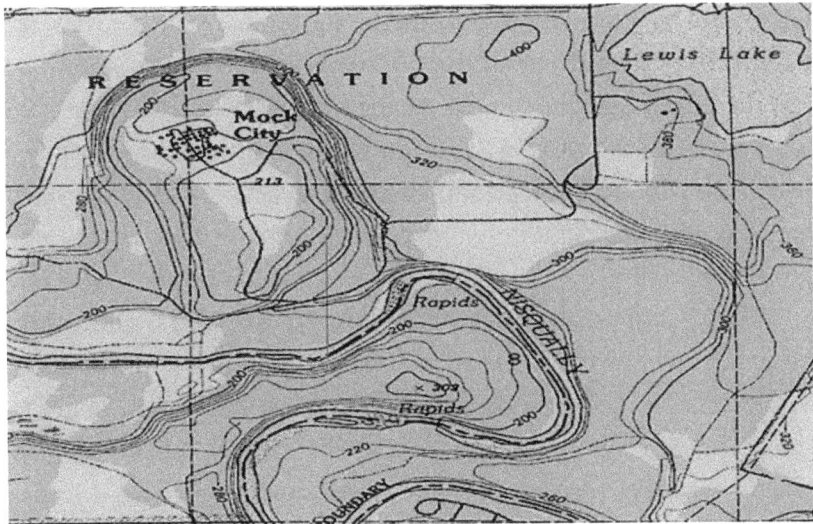

*Nisqually River Floodplain, Fort Lewis, Washington. Each contour
interval represents 20 feet of elevation. Contour intervals that
are scrunched together represent steep elevation changes.*

"Roger that, Sir," he responded as we huddled together to form
a light-blocking human cone around his acetated map, which he
illuminated with a tiny finger light while using a blade of grass
as a pointer. "As you know, Sir, the most prominent terrain feature around here is the Nisqually River. (See map.) If you walk
a hundred meters up the column, you'll come to an incredibly
steep drop-off, which, as you can see on the map, is clearly defined
where all these contour intervals come together. Those bunched-
up contour intervals signify the edge of the Nisqually River floodplain," he explained while tracing the floodplain as it meandered
along both sides of the river.

The next challenge was figuring out our general location on
the Nisqually River floodplain. "We moved a long way tonight,
but the fastest a platoon this size can move in this type of terrain is around two kilometers per hour," Sergeant X explained.
"So even if we had been moving in a straight line at max speed
for the past five hours, there is no way we could have moved far

enough to have wandered outside the training area; otherwise, we would have crossed one of the fire roads that define its boundaries," he emphasized as he traced the outline of the training area with the blade of grass. "As you can see on the map, the only place it's possible to cross the floodplain without crossing any fire roads is inside this area right here," he circled a spot on the map the size of a thumbnail.

"This is where we're at right now, Sir, which is good news because it means the fire road is less than 200 meters north of us. We've actually been paralleling it for the last half hour. To get to the patrol base and get these guys warmed up as quickly as possible, I strongly recommend we move directly north to the fire road and then handrail[10] it for about a kilometer until the fire road turns south. Our patrol base destination is only three hundred meters from that point."

As impressed as I was that he knew where we were, I was even more impressed with the process of elimination he used to figure it out.

"Good job, Sergeant X, now I need you to follow me up to the head of the column so we can get the platoon leader and platoon sergeant squared away." "Roger that Sir, if it's okay with you, I want to go check on my guys first, and I'll jump in behind you when you pass."

"Sounds good," I told him as he hurried back to his squad up ahead. For the first time in many hours, I no longer felt angry, frustrated, cold, and tired. Part of it was knowing where we were and that we weren't that far away from where we needed to go. Mostly, it was my renewed sense of purpose and a platoon full of Rangers who were counting on me to accomplish it.

As mentioned earlier, the main reason that moving through a forty-man column at night in difficult terrain is considered tactically unsound is because it's almost impossible to do so without creating a lot of commotion and noise. With the column now halted, it was going to be even more difficult because most of the men were either sitting on their rucksacks or taking a knee next to a tree. Which meant that their exact locations would be almost impossible for me to see. The only way to prevent accidently stepping or falling on them was to alert each and every one of them as I was coming through.

"XO coming through," I whispered as I waded into the column.

"Who?"

"XO."

"'Exxo?' Are you new?"

"XO coming through, right side of trail."

"Ow."

"Sorry about that."

"That was my leg."

"I couldn't see it; XO coming through."

"What the fuck?"

"Sorry about that."

"Who the hell are you?"

"The XO, who are you?"

"Smith."

"XO coming through."

"Watch out for the two hundred foot drop-off up ahead."

"Thanks man, XO sliding down."

"Who goes there?"

"It's the XO, coming through."

"You got any extra MRE crackers?"

"No, have you seen the platoon leader?"

"Where?"

"He's losing it, Sir; the platoon leader is straight ahead," Sergeant X announced as he merged in behind me.

"XO, what brings you up front?" the platoon leader asked non-chalantly, while sitting on the ground and staring at his map with a red lens flashlight and poncho over his head.

I took a deep breath, *"stay calm; he's new, and maybe he really has no clue."* I began by describing the seriousness of the situation going on behind us. "Of greatest concern," I emphasized, "is that most of the platoon is now dangerously wet-cold." He didn't say anything, so I asked him, "How can you possibly think it makes sense to strip down into your skivvies for any reason during a tactical movement, much less a tactical movement in a rain storm?"

"We actually came up with the idea together," the platoon leader responded indignantly. "Who's we?" I replied. The platoon sergeant cleared his throat to let me know he was standing a few feet off to my right.

"What seems to be your problem with it, Sir?" the platoon sergeant asked with a hint of condescension. The platoon sergeant's unexpected involvement and apparent support not only surprised me, it also made any more discussion concerning the Gore-Tex underwear a dead-end waste to debate. He was both a senior platoon sergeant and a combat vet while I was a lieutenant who hadn't seen combat yet. He could fall back on things like "mental toughness," "discipline," and "experience," while all I could fall back on was that the men and I were freezing our asses off. Compounding my disadvantage was the fact that I was starting to shiver uncontrollably. Which was a timely reminder that what mattered most at that moment was making sure the platoon leader and platoon sergeant knew where we were and how to get the platoon to the patrol base as quickly as possible.

I took a knee to talk directly to the platoon leader. "We can talk more about your Gore-Tex underpants SOP in the after-action review tomorrow morning; right now we need to get to the patrol base and get these guys warmed up as quickly as possible.

Do you have any idea where we are right now?" I asked loudly enough so both of them could hear the question. They responded with thundering silence. "I think I got it figured out, Sir," Sergeant X piped in as he stepped into the light.

"Is that Sergeant X? Hop under my poncho and show me where you think we are," replied the platoon leader.

While Sergeant X oriented the platoon leader and platoon sergeant with his map, I did what I always did when I got wet-cold. I dropped to the ground and started knocking out perfect-form push-ups. The motivation for attempting perfect-form is that they are the hardest to perform and quickest to warm you up.

A few seconds after I finished fifty, the platoon leader announced to everyone in ear range: "Okay, we're all set, we're going to head toward the road and then handrail it the rest of the way to the patrol base." Then he told the platoon sergeant to pass back to the rest of the platoon that we'd start moving again in five minutes. With that warning, Sergeant X and I realized we had to hurry back through the column to grab our rucksacks before the men started moving. I asked him to lead the way this time, which he did.

"Sergeant X coming through."

"Oh geez, not again."

"Sergeant X coming through."

"Huh?"

"Wake up, Ranger."

"You're standing on my hand."

"Sorry about that."

Even though we had only spent a few minutes with the platoon leader and platoon sergeant, it was immediately apparent that the mood of the men had changed from bad to worse. A few seconds after a brutal climb out from the bottom of the 300-foot Nisqually River floodplain, Sergeant X raised his closed fist directly in front of my face. The closed fist is the military hand and arm signal for "freeze." As in "become a statue" of whatever position your body is in when you see it. After a couple of seconds of looking and listening, he whispered, "There's something moving out there." Like cats searching for mice in a field, we both stayed stock still with eyes and ears pointed outward into the forest. And then I heard it too. Something was definitely walking around out there.

The eye is attracted to contrast, so motion stands out. A human figure was walking through the woods about fifty meters from us. The outline of a helmet and weapon confirmed he was a Ranger. *Maybe he's out there taking a piss*, I thought until I heard him start mumbling something. "Is he talking to someone?" I whispered. Sergeant X didn't waste any more time wondering. "Hey, Ranger, this is Sergeant X; are you okay?" He said out loud. No response. Then the mumbling got louder. "Let's go check it out, Sir," Sergeant X said as we both began walking directly toward him, cautiously at first.

"Maybe he lost something," Sergeant X commented as he turned on his red lens flashlight so we could see. "Are you okay, Ranger?" Sergeant X asked as we walked up to him. Still no response. Then the Ranger turned his back on us and broke his silence:

"Popcorn, popcorn, who wants popcorn, get your popcorn while the butter's hot," he screamed at the top of his lungs.

Neither of us needed any more clues. Every Ranger is trained to detect and treat the symptoms of hypothermia. Intense shivering occurs first, followed by unsteadiness in balance or gait, slurred speech, lack of interest, confusion, irritability, and, sometimes, in cases such as this one, hallucinations. Once the symptoms of hypothermia set in, body temperature plummets rapidly. When body temperature goes below eighty-two degrees, the victim can collapse into a coma and die.

"Medic," Sergeant X alerted at the top of his lungs while both of us switched to white lens flashlights. "This is a real-world emergency; everyone go to white lights," I added as we waved our flashlights up and down to help the medic locate us. Within a few seconds, the platoon medic came crashing through the brush and went right to work.

"I'll take one of those popcorns, Ranger," the medic told him in conversational calm. "I just need to get you warmed up first, if that's okay with you?" There was no response from the popcorn salesman, so the medic nodded to me to assist him as we gently sat the Ranger down against the moss-covered trunk of a nearby fir tree. The medic continued talking in the same calming tone while his surgical scissors made quick work of the sopping wet uniform and Gore-Tex underwear. As quickly as the uniform came off, Sergeant X began wrapping the Ranger in a warming cocoon of double down-filled sleeping bags.

While the medic continued to stabilize the patient, Sergeant X organized the stretcher team, and I set up my radio to call our company commander (Captain Gildner) and let him know we had a medical emergency.

"His body temp is ninety-one degrees so I think we got him in time, Sir," the medic updated, "now we need to get him out of this rain and back to the rear as quickly as possible or his body temperature could drop again real quick.[1]"

"Captain Gildner and the battalion ambulance are waiting for us on the fire road," I explained to Sergeant X and the stretcher team.

Even though it was only 200 meters to the fire road, it was 200 meters of dark, damp, deadfall, which made carrying a 200-pound patient on a stretcher just as potentially risky as the hypothermia

---

1 Mild hypothermia = body temperature of 95°F to 90°F, or 35°C to 32°C (shivering, chattering teeth, confusion).
   Moderate hypothermia = body temperature of 90°F to 82°F, or 32°C to 28°C.
   Severe hypothermia = body temperature below 82°F or 28°C.

we were treating him for. We almost dumped him twice in the first fifty meters. Cocooned inside two sleeping bags that were zipped up to his face, we didn't need a thermometer to tell that he was starting to warm up: "I'm not liking this," "you guys are scaring the shit out of me," "Doc, please get me the 'F' off this stretcher and let me walk on my own." As much as we wanted to grant him his wish, the medic reminded us that "he wasn't out of the woods yet," a witty play on words that made everyone laugh and also realize it was the hard right, so we continued to gut it out.

When we got to the road, Captain Gildner was waiting with the ambulance and a couple of senior medics. Hypothermia was, and still is, a big deal in the military in general and the Rangers specifically. A few years earlier, two Ranger School students died of hypothermia, and the incident lived on as both a guiding principle and a lesson for all Army leaders: **"Hypothermia isn't caused by soldiers and bad weather; it's caused by leaders and bad choices. Hypothermia is preventable."**

The mantra was repeated so often and was so intuitively ingrained that in the rare instances, such as this one, where it actually happened, the leaders involved instantly understood the seriousness and potentially career-ending ramifications if they were at fault. A comprehensive report of this incident would have to be completed immediately and sent as a high-priority message for distribution all the way up the Army chain of command.

After the ambulance departed, I thought it was just me and Captain Gildner until I noticed somebody else standing alone in the shadows a couple of hundred meters down the road. His body shape and bearing told me it could only be one person, the battalion commander, Lieutenant Colonel (LTC) John J. Maher.

"Okay, Pete, tell me what happened," Captain Gildner began. I told him about the five-kilometer movement that took us five hours, and that nobody knew what was going on because we didn't get any updates after the first hour. Then I told him about the freezing rain and the order to "don Gore-Tex underneath our uniforms," and about the guys falling down while they tried to take off their pants, and that I stupidly fell down too. Then I told him about Sergeant X finding me, and what he told me about his guys' chattering teeth, and how he figured out where we were and squared away the platoon leader.

Even though it was too dark for me to see his face, I could tell Captain Gildner was beside himself with anger. Whenever he got really mad, he'd start walking around in little circles while he talked. "This is un-freaking believable, Pete," he lamented as he logged his first lap, "what happened when you went to talk to him about the freaking Gore-Tex underwear?" I told him about my conversation with the platoon leader and platoon sergeant and that Sergeant X had already figured out where we were, so instead of wasting more time trying to talk sense into them, we focused the conversation on getting them back on track.

When I got to the part about me and Sergeant X heading back to grab our rucksacks before the platoon started moving, he stopped me. "So instead of calling me on the radio to tell me what happened, you were heading to the back of the column to continue following along like some sort of mindless lemming for the rest of the night?" Before I could say "Yes, Sir," and acknowledge my now painfully apparent F-up, he continued, "For all you know, Pete, they might have changed their minds and decided to head down to the Nisqually River to conduct a poncho-raft river crossing." Hilarious in hindsight, but he didn't laugh, so I didn't either.

"I know you were probably trying to be respectful and let the platoon work things out on its own, but this is a good example of a situation where your responsibility as a leader overrides your responsibility to let somebody learn. There's no punishment for bouncing what you're seeing or hearing off someone else, which is the main reason you're humping that twenty-pound brick [the radio] around on your back. If you had called me, I would have told you this is the third time he's done this in the last six months, and then I would have told you to take over the platoon on the spot.

"When were you planning on telling me about this, Pete?"

"I didn't want to bother you in the middle of the night, Sir, so I figured I'd tell you about it tomorrow morning," I admitted, once again fully aware of how stupid it now sounded. He took a deep breath. "What's my job, Pete?"

"To command the company?"

"Nope, that's my job description. My job is the same as yours; it's to take care of the men in this company. Taking care of the men isn't some touchy-feely thing that means we coddle them— that's not the purpose they came here to accomplish. **Taking care of the men means making good choices that set the conditions for them to succeed. That's how we create a healthy leadership climate.**"

Leadership climate was a topic that leaders in our battalion talked about a lot. Yet it wasn't Captain Gildner's words that taught me, and the rest of the leaders in the company, how important it was, it was his actions and his choices.

He told us to "think of a leadership climate the way we think of an actual climate; you can see and feel the effects of the climate all around you." When a leadership climate is healthy, it sets the conditions for everyone and everything under it to grow, progress, and succeed. Everyone wants to live under a healthy leadership climate, so when they find one they like and trust, they put down roots and work together as a team to make it better for the future. But when the leadership climate is unhealthy, individual growth is stunted, the people who live under it are miserable, and instead of focusing on the mission and how to make things better, all they think about is going somewhere else or getting out.

Even as I stood there in the middle of the road, shivering like a sorry sack of dog excrement, I knew he was teaching me an important lesson. I already knew and appreciated him as the best company commander in the battalion; over time, and across many other leadership experiences, I would come to realize he was a key part of the best leadership climate I ever had the opportunity to be part of. And that's where the other guy on the road that night came in.

After Captain Gildner finished talking, he told me, "The battalion commander wants to have a word with you; he's been standing out here for three hours waiting for the platoon to arrive, so he may not be in the best of moods."

As I hurried down the road to meet with him, I glanced at my watch, it was 0430. This was a company training exercise on a Tuesday night, so why would the battalion commander stay out here this late and in this weather? Even though I recognized there was still a possibility I might get kicked out of Ranger Battalion for what happened, I felt no trepidation as I approached him. I didn't fear my battalion commander. I respected him. There's a big difference between being reprimanded by someone you trust and respect, and someone you don't. Trust and respect aren't earned in a single transaction; instead, they are built over time by the aggregate of all of our actions and choices. In all the time I worked for him, he was always logical, always fair, and everything I ever remember him telling me made sense. So whatever he decided, I had no doubt I'd respect his decision.

Born and raised in Georgia, he was 6′4″ tall and spoke with a slow Southern drawl and a deep, calming voice that he used to impart his signature wit and homespun wisdom whenever he talked.

He began by telling me I was lucky to have a company commander who allowed me to learn, and then reiterated that letting your people learn has limits. "A good rule of thumb that has always worked for me is anytime the men are senselessly suffering, it's time to intercede." He went on to tell me that although he was ultimately responsible for the leadership climate in the battalion, he couldn't create or maintain a healthy leadership climate without his subordinate leaders at every level.

"I have a responsibility to understand how my subordinate leaders interact with their people. That relationship is the glue that holds every great combat team together. I wasn't planning on watching Captain Gildner and you interact tonight, but what I saw reinforced what I already knew, that you guys make a great combat team. I was hoping to watch the platoon leader and platoon sergeant set up their patrol base, and, who knows, maybe they might have set up the best patrol base I'd ever seen, and I would have walked away thinking I saw a combat-ready platoon. I wouldn't have had any way of knowing about the five-hour treasure hunt they led their men on before they got here, nor would I have known that the reason the men all looked overweight was because they were wearing their Gore-Tex rain suits as underwear." Once again, I wanted to laugh, but he kept talking, so I kept silent.

"That's why I depend on leaders at all levels like Captain Gildner, Sergeant X, and you. I can't monitor the command climate where you're at because I don't live there. **A leadership team has to work together by always sharing what they see."** He then told me to think of myself as one of his climate monitoring stations. **"If you see a tornado touch down, you have a responsibility to share that information so the rest of the leadership team can take action and make sure none of our people get hurt by it.** By not reporting the tornado, you're allowing it to destroy the next town in its path and the next one after that. **Leaders are most often thought of individually, yet the real power of leaders is collective. Every leader contributes to the creation of a healthy leadership climate, and every leader is responsible for maintaining it."**

He summed up everything he told me with his signature homespun Southern wisdom: **"The only way a leader can see his own ass is to have someone he trusts tell him what it looks like."** This time I laughed out loud, and he did too. Then he told me to **"remember what happened out here tonight, learn from it, and make sure it never happens again."** "Roger that, Sir," I responded with gratitude, relief, and newly learned appreciation for the importance of a healthy leadership climate. **"I'll never allow it to happen again, Sir, you have my word."**

A Company Commander, Captain Gray Gildner, in front; at the time, I was A Company executive officer (XO), and I'm standing behind him over his right shoulder.

CHAPTER 2

What a Healthy Leadership Climate Looks Like

I was very fortunate throughout the early part of my career to work for so many common sense leaders. The military, at its best, is a sprawling leadership academy. After a couple years of attendance, you have seen and heard about the importance of taking care of your people and always doing the right thing for them so many times, over and over, that it sticks with you for life.

Life lessons take time to learn. Only while writing about and reflecting on this period of my life did I comprehend both the immediate and long-term impact of common-sense leaders and the healthy leadership climates they create:

- **What's the purpose of all leaders?** To take care of their people.

- **How do common-sense leaders take care of the people they lead?** Taking care of our people means using common sense to make good decisions and solve complex problems that set the conditions for our people to succeed.

- **What is a leadership climate?** Think of a leadership climate the way you think of an actual climate; you can see and feel the effects of the climate all around you. When a leadership climate is healthy, it sets the conditions for everyone and everything under it to grow, progress, and succeed. Everyone wants to live under a healthy leadership climate, so when they find one they like and trust, they put down roots and work together as a team to make it better for the future.

- **How do common sense leaders create a healthy leadership climate?** A healthy leadership climate is one characterized by a high level of trust, respect, and loyalty. A healthy leadership climate emerges over time via the sum total of sensible choices made by all leaders within the climate system. Every leader contributes to the creation of a healthy leadership climate, and every leader is responsible for maintaining it. Healthy leadership climates reveal that the true organizational power of leaders is collective.

- **What patterns are pervasive in a healthy leadership climate?**
  - ▶ A common sense of shared purpose and direction amongst all members on all missions all of the time.
  - ▶ Collaboration across boundaries is common and encouraged.
  - ▶ Communication is free-flowing and omnidirectional.
  - ▶ Reciprocity is the currency of all interactions and the foundational governing principle upon which trust and building best teams are based ("you get what you give").
  - ▶ Leaders say out loud the logic of why all orders, plans, or directives do or don't make sense.[11]
  - ▶ The logic of why is always present and/or always available on request.
  - ▶ A common sense amongst all members of what does and doesn't make sense.

- ► Freedom to make sensible choices on the spot without permission.[12]
- ► Freedom of speech to speak up, out, or with all leaders.
- ► Standard operating procedures make sense to those who have to follow them.
- ► There is no favoritism or cliques. Character, competence, courage, and common sense are prized and prioritized.
- ► There are no barriers or restrictions against attacking senselessness and bureaucratic aberrations in the system.[13]
- ► Healthy leadership climates often contain an element of good humor. Team members can laugh about what they are doing and share a good laugh with their leaders.[14]

The immediate impact of a healthy leadership climate is exemplified by the common sense leaders it attracts. The long-term impact emerges over time from the sum total of choices those common-sense leaders make, the behaviors they model, and the knowledge they pass on to those who are fortunate enough to work with, and for them.

Common Sense of Purpose and Direction
+ Time + Feedback = Culture

The foundation of a healthy leadership climate is leadership in general and senior leadership in particular. Healthy leadership climates start at the top, in this case with our Ranger Battalion Commander, Lieutenant Colonel John J. Maher. What follows is a small sample of the timeless leadership knowledge and principles that he and two of his key subordinate leaders taught me and that I still use and pass on to this day.

Battalion Commander, Lieutenant Colonel John J. Maher:

"The best form of welfare for our troops, when we're not deployed, is getting them home in time to eat dinner with their families."

"Unless we are preparing for a mission, the only car I should see in the parking lot past six p.m. is mine. If you're watching and waiting for my car to leave, you're violating two of our standards because you're telling your people that hanging around for no reason is okay and you're preventing your people from making it home in time to eat dinner with their families."

"Friendship is one of the most misunderstood and under-appreciated aspects of leadership. Some say you can't be friends with your people; I say, have you ever had a friend that you didn't trust and respect? So how can you create a leadership environment based on trust and respect if you don't treat your people with trust and respect? Always remember, friends don't let friends drive drunk, and when a friend does drive drunk, real friends take them to task for it."

"How a leader communicates on the radio is in many cases more important than what the leader communicates. Always take a deep breath before you talk, and speak calmly no matter what's going on around you." Tone of voice carries far more weight with our brains than our words (the brain takes 10% of message meaning from words and 90% from the tone of voice and body language used to deliver the words). Your tone of voice is what is queuing your people. Airline pilots and flight attendants are taught this in their version of basic training. "If your people hear you panicking or getting angry on the radio, they are more likely to panic and get angry themselves." Actions control thoughts. Take action and control your emotional thoughts by speaking calmly. **When we act calm, we calm the way we act.**

Battalion Command Sergeant Major Leon-Guerro:

As the battalion commander's right-hand man, Sergeant Major L-G as we called him, was a muscle-bound Guamanian whose booming voice, powerful handshake, and gap-toothed smile were always a welcome addition to any situation he walked up on. **The most foundational lesson he taught us was the guiding principle, "Don't be in a hurry to die." He was constantly priming our minds with its all-purpose importance.** Here's how he explained it to me:

"Impulsiveness is one of the biggest killers on the battlefield. Warriors are aggressive by nature, but acting without thinking instead of understanding the situation first could be the last mistake you ever make. It's what happened at LZ X-Ray."[15] One of the biggest and most famous battles of the Vietnam War began when a platoon leader—who had been in the Army less than a year—spotted a single North Vietnamese soldier standing on the edge of the jungle clearing as his helicopter was about to land. The instant the helicopter touched down, the platoon leader jumped out and bolted after the enemy soldier without telling the rest of his platoon where he was going or why. All twenty-nine members of the platoon followed their leader—as they were trained to do—running after him into the jungle until they couldn't run anymore. When the platoon leader finally paused to catch his breath, he and his men discovered they were surrounded. The battle of LZ X-Ray was actually the battle to rescue the wayward platoon and their impulsive platoon leader. Twenty-six hours later, nine men, including the platoon leader, were dead. Thirteen others were severely injured. Only seven were able to walk out of the jungle unscathed.

Before SGM Leon-Guerro taught me about the perils of impulsiveness, I was that platoon leader.[16] Fortunately for me (and my men), I spent my first four years in the military waging war against pretend enemies who fired blanks instead of bullets. By passing on and teaching us the lessons learned from LZ X-Ray and the wayward platoon, we learned about the fine line between aggressiveness and impulsiveness. As a result, SGM Leon-Guerro changed the way I thought, acted, and led for the rest of my career. Whether I was thinking of running to the sounds of unknown guns, or running off a helicopter on a dark landing zone, or running into an uncleared building, conscious awareness of the guiding principle, **"Don't be in a hurry to die," and the logic of why it makes sense—"impulsiveness leads people to their deaths"**— saved my own life and the lives of those around me too many times to count. For fellow first responders, especially, prime your people's minds before any type of mission by saying it out loud and repeating it.

Battalion XO, Major Bill Leszczynski:

Major Leszczynski was a deep-thinking, super-dedicated and disciplined student of maneuver warfare. Whether we were talking about the pluses and minuses of leaving a stay-behind ambush in the jungles of Panama or the latest breakthroughs in sports nutrition and human performance, he always seemed to have an insightful anecdote or germane fact to add to the discussion as if it had been sitting on the tip of his tongue from the first moment we started talking.

He carried a pocket notebook with him wherever he went and was constantly pausing in the middle of conversations to write down his thoughts and insights as soon as they popped up or into his head.

I bought my first pocket notebook in 1988 and would go on to write down most everything I learned while carrying it with me for the rest of my military career. I still carry one wherever I go to this day (this paragraph came from one of those notebooks).

Finally, many of my peers in the 2nd Ranger Battalion who were lieutenants and captains at that time went on to lead some of the most successful missions and units during combat operations in both Afghanistan and Iraq. Many of us have stayed friends throughout our military careers and are still friends today.

The lessons I learned from these leaders and the leadership climate they created became permanently etched inside my brain. Over time, and with feedback, they formed the foundational knowledge upon which I led and organized for the rest of my leadership life.

Common sense leadership and healthy leadership climates go hand-in-hand. Over time, they create great organizational cultures and great organizational success. **How common sense leaders create a healthy leadership climate is learned. Pass it on. It's the Common Sense Way.**

PART II

WHAT HAPPENED TO PAT TILLMAN AND HIS PLATOON IN AFGHANISTAN

CHAPTER 3

How Part II
is Organized

The story that follows isn't about Pat Tillman per se; it's about what happened to Pat Tillman and his platoon in Afghanistan and **what "we the people" can learn from it**. The declassified[17] versions of the four official investigations into the death of Pat Tillman totaled over 3,500 pages of statements, findings, pictures, videos, and maps, all of which I used to put together the chronology of events and the lessons that follow.

The question I continually asked myself while researching, reliving, and reflecting on what happened was, how can the rest of us learn from this tragic situation to further our own knowledge and make fellow first responders safer in the future? A professional body should be able to examine its past performance openly and honestly, admit shortcomings candidly, and take action to rectify them. This was the spirit in which I researched and wrote these pages.

Congress mandated the declassification of the four investigations so that their findings could be handed over to the Tillman family upon completion. Each additional investigation was launched due to deficiencies found with the investigations that preceded it.[18] To declassify the documents, the military redacted most of the leaders' names as well as some ancillary information that was, at the time (2004–2007), deemed as sensitive. The redactions of names didn't matter to me because I had no intention of using them. One of the reasons investigations such as this one never see the light of learning eyes is for this very reason—the names of the participants. The higher their rank, the higher the degree of scrutiny and sensitivity applied by the redactors.

Most of the Rangers' actual names have already been used in numerous magazines, books, and online news sources. Some of these articles have focused out-of-context criticism on the men of the platoon, which only makes it harder for the psychological wounds to heal. Additionally, some of the individuals who contributed to this manuscript are still government employees who did not want their names mentioned for fear of retribution. As a result, I have chosen to anonymize the names of the individual Rangers who were involved in this incident.

Trying to make sense of heavily redacted documents without key nuggets of context such as names, locations, and call signs can be quite difficult. This is where I had a huge advantage. In the spring of 2002—two years before this incident happened—I lived in, walked through, and drove around the same villages, terrain, and roads as the Rangers of 2nd Platoon did. I had talked to and interacted with the same Afghan tribes, warlords, and tribal elders who controlled the people and the places the Rangers were operating in at the time. Finally, as mentioned earlier, I served in the same unit (A Company, 2nd Battalion, 75th Ranger Regiment) with some of the same men and fought against the same amorphous enemy both before and after this event happened.

Getting information from individuals who were part of the military investigation process presented its own set of challenges. When I began my research, Mary Tillman shared stories with me about how she was treated. In short, she was yelled at, mocked, hung up on, and ignored by government officials. Many of them were indignant that she still wasn't satisfied with the explanations she received from the four official investigations. "What more do you want?" "There's nothing else here." "Why can't you accept the facts and leave well enough alone." Initially, I was shocked that anyone would treat the mother of a fallen soldier that way. It didn't take long before I was experiencing the same thing.

One of the lead investigators was an officer whom I knew and had served with during the early days and weeks on the ground in Afghanistan. I had a solid working relationship with him, so when I sent him an email asking if he "had time to connect," he told me to call him the next day. When I called, he was boarding an airplane, and I could hear the stewards in the background asking the passengers to turn off their cell phones. "Do you want me to call back later?" I asked. "No, tell me what you're up to these days."

I quickly explained that I was doing research and writing about what happened to Pat Tillman and his platoon in Afghanistan. Before I could add context, he cut me off and began screaming: "Why the hell do you want to do that, there's nothing else to learn, it's all in the reports, I don't ever want to talk about that event again." He was still talking when his phone cut-off so I wasn't sure if he hung up or if the Wi-fii was shut down. I later learned that he was forced to retire after he received a letter of reprimand for the way he conducted the third investigation, specifically, for "incorrectly characterizing Corporal Tillman's actions in describing why he should be awarded a Silver Star." *Since when do the investigators get letters of reprimand?* I wondered. He never answered my calls or emails after that conversation.

Even though it's always difficult to understand and make sense of events that happened in the past, this particular incident distinguishes itself with regards to historical scrutiny. I'm not aware of any other modern day battlefield event that was investigated so thoroughly and memorialized with such an abundance of information. The four investigations produced redundant and overlapping documentation on details such as terrain, weather, weapon systems, time of day and year, as well as personal accounts from almost every Ranger involved (some Rangers were interviewed seven separate times and asked many of the same questions each time). Access to the above information enabled me to credibly compare, contrast, and cross-reference it to create a coherent narrative that makes sense of what happened.

Despite being well suited by experience to make sense of the sequence of events that follow, I can't claim flawless exactitude for this account any more than the military investigators and writers who have written about it before me. The main reason is constantly reinforced within the 3,500+ pages of the investigative documents: most everything we know about this incident comes from personal accounts in the form of written statements and answers to the investigators' questions during the weeks, months, and years after it happened. No two people make sense of an experience exactly the same way. Time, location, life experiences, state of mind, perspective, mood, and memory are just a few of the many factors that effect the way we perceive and recall the details of any life experience.

The descriptions of events below are organized chronologically. The narrative includes comments and statements that were taken directly from the official investigation documents, and/or are quoted directly from the sources. As you read the statements, you will notice that some sound a bit jumbled or incoherent. This is reflective of the fact that people don't speak the same way they write and that the Rangers' spoken answers to the investigators' questions were recorded and translated into written words using speech recognition software. In some cases, the quotes contain military acronyms or jargon that I clarify with explanations in brackets; otherwise, you are reading the quotes exactly as they were recorded in the redacted investigation documents.

To the extent possible, I ran everything by a group of three former Rangers who were on the ground in Iraq and/or Afghanistan during the time of the events. I also conducted over 100 interviews with 26 Rangers from the 2nd Ranger Battalion, who were eager to tell their stories, and overwhelmingly appreciative that someone was asking them about what they experienced and listening to what they had to say. Most of my sources asked to remain anonymous—a pervasive pattern I was already familiar with and which I fully understood and supported.

Finally, as noted at the beginning of the first chapter, I delayed publication of this story because two of the three Ranger non-commissioned officers (NCOs) who agreed to review and contribute to the manuscript were out of the country at the time (2018–2021). It turned out to be one of the best decisions I could have made. Their input provides context-specific first-hand knowledge of conversations and events: from what was going on at Baghdad International Airport, to what was going on in the Rangers' Forward Operating Base in Afghanistan, to information about the actual firefight that was not captured in any of the four official investigations.

Sergeant Major (ret.) Ted Kennedy deserves special mention and thanks for the advice and counsel he has provided. As you'll learn in the pages that follow, he spent 17 years in the 2nd Ranger Battalion and was the first sergeant of A company when the Tillman brothers arrived and while the battalion was operating out of Baghdad International Airport in 2003. He is quoted verbatim throughout the book, and his insights were instrumental in helping to shape the narrative and the lessons (in bold) into something that all leaders could understand, learn from, and, most importantly, use when similar situations arise in the future.

To see high resolution color versions of the photo's that
follow, as well as full length videos of the drive through
the canyon and Serial #1's movement to the spur, see:

www.firefight.commonsenseway.com

CHAPTER 4

The Intersection of 2/75 in Iraq

24 April 2004, Pentagon, Northern Virginia:

"Flash report" is the term the U.S. Military uses to describe high-priority messages. Flash designation is reserved for messages categorized at the highest level of urgency. Department of Defense guidance for flash reports states: "just the bottom line facts as the sender knows them with clarifications to follow." Brevity is a mandatory requirement for flash report messages. Those that receive flash messages are instructed to read and forward them as fast as possible, ahead of all other messages. It was in this context that I read the flash report[19] that was displayed in front of me on my desktop computer.

Topic: Corporal Patrick Daniel Tillman killed in action (KIA) during enemy ambush from gunshot wound to the head. One Afghan coalition partner also KIA. At least two other U.S. personnel wounded in action (WIA).

Location: Vicinity Spera District, Khowst Province, Afghanistan.

Unit in Contact: 2nd Platoon, Alpha Company, 2nd Battalion, 75th Ranger Regiment, Fort Lewis, Washington.

Enemy Forces: Unknown.

Enemy KIA, WIA, and EPW[20]: None.

I always try to remember the first thing that comes to my mind when I experience or learn something new. The best way to remember and learn from our first thoughts is to say them out loud: *Sounds like friendly fire* I mouthed to myself.

First reports are almost always inaccurate, incomplete, or both. This one, I was hoping, would be no different. I kept reading the message over and over. A couple of key details jumped out as prominent.

The first was the "gunshot wound to the head." A Kevlar helmet covers most of the wearer's head, and Rangers wear their Kevlar helmets at all times, on all missions. So either the enemy got really lucky or they were really close. Most enemy we had encountered in those first few years in Afghanistan (2001–2002) carried iron-sight AK-47s or one of its numerous variants. None of which are very accurate over 200 meters. Some foreign fighters we had encountered in the first few months were equipped with various versions of the Soviet SVD-Dragunov sniper rifle; however, most of the foreign fighters had fled to Pakistan in the spring of 2002. Up to this point in the war (2004), neither the foreign fighters nor the Taliban had demonstrated the capacity or the intent to execute a deliberate close-range ambush against a heavily armed, highly trained, and highly disciplined unit such as the Rangers. Which makes sneaking up on and/or overrunning them even more improbable.

The second detail was related to the first: No enemy KIA, WIA, or EPW. Even if an enemy force had somehow managed to get close enough to shoot a Ranger like Corporal Tillman in the head, it was highly unlikely they could have gotten away without some of them getting killed, wounded, or captured (KIA, WIA, or EPW). To be clear, my "friendly fire" first thoughts were neither an opinion nor a prediction. I was fully aware that I didn't know any firsthand facts or details surrounding the incident and was hopeful that I'd be proven wrong.

Yet it wasn't just the lack of enemy acumen or the lack of enemy KIA/WIA/EPWs, nor was it my firsthand knowledge of the combat power and proficiency of the Rangers that was driving my skepticism. Instead, my skepticism was driven by recognition of one of the least talked about yet most frequently occurring patterns of modern-day combat: **friendly fire**.

At this point in my career (2004), I had been to combat in Panama, Somalia, Bosnia, Afghanistan, and Iraq, and I had directly or indirectly experienced friendly fire incidents in: Panama, Somalia, Bosnia, Afghanistan, and Iraq. **When it comes to combat and friendly fire, experience in one is usually proportional to experience with the other.**

War/Campaign	% Friendly Fire Casualties*
World War II	21%
Korea	18%
Viet Nam	39%
Panama	.08%
Persian Gulf	52%
Afghanistan	13%
Iraq	41%

*Source: American War Library, Friendly Fire Notebook. Percentage of casualties includes both fatal and non-fatal injuries. Numbers reflect U.S. military only and do not include the deaths of friendly allies or civilians.

"Expert" has the same Latin root as "experience." Aldous Huxley, writing in his book *Texts and Pretexts*, said: "Experience is a matter of sensibility and intuition, of seeing and hearing the significant things, of paying attention at the right moments, of understanding and learning from it." **Experience in combat or with friendly fire doesn't do us any good unless we learn from it.**

My first friendly fire experience was in Panama (1989–90), though I didn't understand the fire we took and returned was friendly until four years later. Once I understood what happened, I was able to learn from it. Even though no one was injured, it made me reexamine every firefight in which I took part or was indirectly involved. Although all of my friendly fire experiences were contextually different, **there was one common pattern that always remained the same. After it happened, no one wanted to say the words "friendly fire" out loud. Especially when someone died.**[21]

Thus, when I combined what I had just read in the flash report with all the friendly fire patterns I had learned from my past experiences, it was my belief that the fire that killed Pat Tillman was far more likely to be friendly than enemy.

What really happened to Pat Tillman? I wondered as my mind raced backward in search of answers. Although Pat Tillman and I had served in the Rangers at different times, our career paths intersected at the same place. Known to those who have passed through it as 2/75 (pronounced: two-seven-five), the 2nd Battalion of the 75th Ranger Regiment is one of three Ranger Battalions (1st, 2nd, and 3rd). The 1st and 3rd Ranger Battalions are based in Georgia, while the 2nd is based in Fort Lewis, Washington.

I served two separate tours of duty as an officer in the 2nd Ranger Battalion. I first passed through as a lieutenant between 1985 and 1989. As previously mentioned, for two of those years (1987–1989) I served as the executive officer (XO) of Alpha Company, which is the same company that Pat Tillman's 2nd Platoon was part of, and the same company and platoon I walked the woods of western Washington with.

The last time I had had any interaction with Rangers from 2/75 was almost one year to the day earlier on 30 April 2003, during combat operations in Iraq. That particular night is deeply etched

OVERVIEW
Where the Regiment is Located

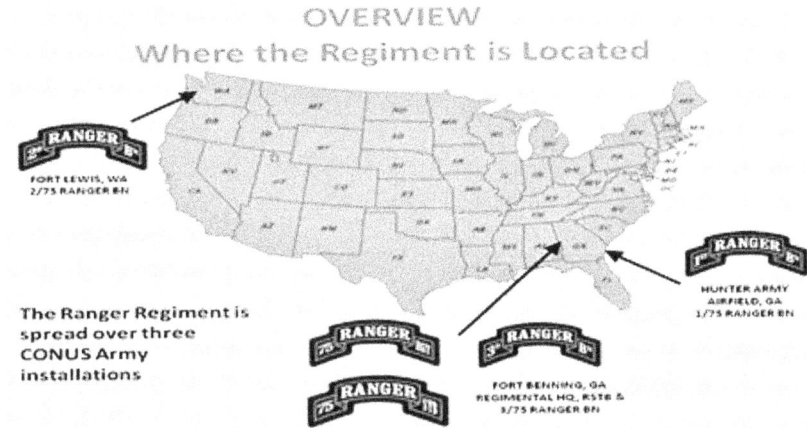

RANGER

FORT LEWIS, WA
2/75 RANGER BN

RANGER

HUNTER ARMY
AIRFIELD, GA
1/75 RANGER BN

The Ranger Regiment is
spread over three
CONUS Army
installations

RANGER

RANGER

RANGER

RANGER

FORT BENNING, GA
REGIMENTAL HQ, RSTB &
3/75 RANGER BN

in my memory as the night our unit experienced the highest number of gunshot-induced casualties (four wounded) on any mission up to that point in Afghanistan or Iraq.[22] Some of my very best men were seriously wounded on that mission.

Like many other U.S. military units on the ground during the first few days and weeks after the Saddam government fell, we were using the Baghdad International Airport as our secure staging base. The airport's perimeter fences and unobstructed stand-off distances[23] surrounding the hangers and terminals provided us with an isolated and secure location from which we could launch 24/7 capture-operations in and around the city.

Our temporary tactical operations center (TOC) was located in what used to be the International Terminal. The departure gate doors provided us with rapid and secure access to our vehicles and helicopters, from which we could launch forces on a moment's notice whenever we had actionable intelligence.

When the medevac helicopters landed that night with the four wounded Unit operators onboard, we didn't have enough people to carry all four stretchers, so about a dozen Rangers who were also operating out of Baghdad International Airport volunteered

Baghdad International Airport (2003)

to help carry the stretchers into the triage area where our medical augmentation personnel were waiting to treat them. I found out they were from the 2nd Ranger Battalion when I thanked them.

After the Rangers left, one of my wounded men told me, "We could have used a couple of those Rangers around the target, Panther; we were a little thin on our isolation/security force." "Got it, Chris," I reassured, just before the anesthesiologist put him under. Later that night, I called our higher headquarters and requested a platoon of Rangers to augment our assault teams whenever needed on future operations.

The next night (1 May), I was standing behind the Delta ticket counter talking to one of my teams on the radio when I noticed a Ranger walk through the terminal doors. I heard him ask one of my guys if he knew "where Colonel Blaber was," and instantly recognized the voice of my old friend Sergeant X. Sergeant X and I had worked together on both of my tours in the 2nd Ranger Battalion, most recently during my second tour from 1996–1998 while I was serving as the battalion operations officer/S-3.

Even though it had been five years since I'd left the Rangers, and five years since I'd last seen Sergeant X, my excitement to see him was 100% operational. I thought his presence meant my request for Ranger augmenters had been approved and that he was the lead Ranger coming over to coordinate. I greeted him with a smile that he uncharacteristically did not reciprocate. "Sir, can I talk to you in private?" was all he said. "Of course," I replied as I grabbed my long gun and motioned for him to follow me outside through the departure gate doors into the jet-black stillness of the no electricity, no ambient light, Baghdad night.

Sergeant X didn't waste any time on small talk. "Sir, I know you're not in the Rangers anymore and you're obviously super busy, but honestly, I didn't know who else to talk to," he began. "No worries, Sergeant X, what's on your mind?" I asked him as we began walking and talking along the inside edge of the inactive 10,000-foot-long runway. "It's the leadership climate in the battalion, Sir. While you and your guys are conducting missions around the clock and taking the fight to the enemy like we're all supposed to be doing, we've been standing around outside our tents in the middle of the night with our assault gear laid out on ponchos while waiting to get our canteen cups inspected for cleanliness. If one of our guys doesn't do well in the inspection, his leaders have to pick up trash around the airfield the next day."

Formal inspections of canteen cups sounded like something out of the post-Vietnam military of the 1970s; leaders being forced to pick up trash around an enemy airfield during combat operations is something that's almost impossible to comprehend. It might have been funny if it had happened back in the U.S. while training, but this wasn't training, this was combat; and we were fighting a real enemy.

"The worst thing about it is that we have no time to do the precombat tasks that we all know we should be doing to ensure we keep our guys alive; like making sure all our weapons and target acquisition devices are zeroed, or practicing our movement-to-contact and immediate-action drills, or working with the medics to rehearse our advanced trauma and life-saving techniques. And, Sir, I'm not coming to you just because I don't like inspections; I'm telling you all this because **the leadership climate is so toxic, and everyone is so stressed out about it that I'm concerned something really bad is going to happen.**"

As he continued to describe examples of petty senselessness that he and his fellow Rangers were going through, my mind toggled back through the faces of some of the men he mentioned that I had served with. By seeing the world through their individual eyes, I could feel their pain. Even though I didn't know any of the leaders in the chain of command he was referring to, that didn't mean I couldn't relate to what he and his fellow Rangers were going through.

I served a total of six years in the 2nd Ranger Battalion, and the leadership climate during all six was either very good or the very best I had ever worked under. Yet I was also aware of at least two other periods of time during the years I wasn't assigned to 2nd Ranger Battalion that had had leadership climates described to me as both tyrannical and toxic.

Most Ranger NCOs I had worked with rarely, if ever, complained about anything. They were so used to "sucking up the senselessness" in the name of selfless service that you literally had to coerce and cajole them to get them to tell you what, if anything, was actually bothering them.

Sergeant X was what one of my former Ranger commanders referred to as "hard as woodpecker lips." A perfect example of his woodpecker hardiness occurred on one of our monthly twenty-five-mile road-marches. I was walking behind Sergeant X and noticed he was limping and struggling to keep pace, which was unusual because he was Olympic-athlete fit and normally the pacesetter. I walked alongside him and asked if he was OK. After a couple seconds of silence, he uttered, "I'm fine, Sir. My foot is

F-ed up, but I'm OK." "Do you want the medic to take a look at it?" I asked. He shook his head no and swallowed hard before adding, "Maybe after we finish." I radioed ahead and asked our world-class physician assistant (Chief Egan) to meet us at the finish line, where he wound up having to order Sergeant X to sit down so he could check him out. It turned out that Sergeant X had broken his foot on a parachute jump three days earlier.

As I thought about what Sergeant X had just told me regarding the current leadership climate in Baghdad, I realized how much pain and frustration he and his fellow Rangers must have already sucked up to motivate him to jump across command lines and ask for help from an officer that was no longer even in the Ranger regiment. I should also add, that based on what had happened to my guys less than twenty-four hours earlier, I had no uncertainty about the seriousness of this type of situation in a combat zone. **Senseless leadership in a combat zone is as deadly serious as it gets.**

"Have you talked to any of the leaders in your chain of command about this?" I asked him. "Yes, Sir, quite a few of us have tried, but this command has made it known that there is no tolerance for dissent. They hand out career-ending Article 15s[24] and letters of reprimand like parking tickets. Most of the senior NCOs and company grade officers[25] want to stand up for their men and do the right thing, but they also have families to feed and can't afford to get kicked out of the battalion and have their careers ruined by a letter of reprimand or a below-average evaluation report, which only requires the stroke of a pen. I guess you can say we've resorted to doing what Rangers always do: sucking it up and driving on."

"What can I do to help?" I asked with both sincerity and uncertainty while we continued to walk and talk.

"I'm not sure, Sir…you've already helped me just by listening…I apologize again for bothering you with something like this, but I had no idea who else to talk to. I know you know a lot of people in our higher headquarters, so maybe you could talk to one of them and get someone to come and take a look for themselves."

Just then my handheld radio crackled to life: "We need you back at the TOC ASAP." "Roger. On my way," I replied.

I summed up by giving Sergeant X my word that I'd do everything I could to find the best person to talk to. I also told him I would make sure that whoever I talked to understood how serious the situation was, as well as how urgent it was to do something about it quickly. I had to add that I couldn't make any promises on timing because I hadn't seen anyone from our higher headquarters for weeks and we (the Unit) were scheduled to move out of the Baghdad International Airport to a safe house in the next couple of days.

The next morning, while walking to the chow tent, I bumped into two senior sergeants from the 2nd Ranger Battalion. Like Sergeant X, I knew and had worked with both of them five years earlier. After a couple minutes of small talk, they looked around to make sure no one was watching, and one of them asked, "Can we get off the beaten path and talk somewhere private?" "Of course," I responded as we ducked into the abandoned luggage terminal, sat down on the conveyor belt, and started to talk.

Over the next hour, they took turns telling me many of the same things that Sergeant X had told me: "Daily formations for no reason, daily 'police calls' to pick up trash, spending hours trying to make random stuff look neat, and highly dysfunctional cot layouts." They also shared some additional details that Sergeant X hadn't. "The issue that causes the most angst across the battalion is micromanagement and lack of trust in subordinate leaders' recommendations. They refuse to listen to what the NCOs and company grade officers are trying to tell them about what is going on outside the wire and what we should be doing to prepare for it. The only feedback we ever receive is negative, even after we come back from successful missions. **We are constantly being reprimanded**

because someone was out of uniform, or had their hands in their pockets, and the like."

The final incident they told me about was reflective of what they described as initiative-crushing micromanagement. After returning from a successful all-night mission, a platoon of Rangers were waiting for helicopters to take them back to Baghdad International Airport. They were in a secure area of the city known as the Green Zone. At the time, the Green Zone was more secure than Baghdad International Airport. After the chain of command informed them that the helicopters were delayed for two hours, the platoon leader and platoon sergeant told the men they could use the time to get some well-earned and much-needed sleep so they would be ready to go again if another mission popped up.

A few minutes later, the platoon sergeant received a call on the radio from the battalion sergeant major, who ripped into him for allowing his men to fall asleep with their helmets and body armor off. He was told to wake them up immediately, have them put their helmets and body armor back on, and then pick up trash until the helicopters arrived. It turned out the battalion chain of command had used a Predator (unmanned aerial vehicle or UAV) to check on the platoon and their adherence to the battalion's uniform policy.

Note: After the incident, the term UAV took on a new meaning to the Rangers of 2/75. Instead of standing for "unmanned aerial vehicle," from that day forward, whenever the men were about to take off their helmets they would check the sky to make sure they weren't being watched by the "uniform adherence vehicle."

"Once an incident, twice a coincident, three times and it's a pattern." After the two senior sergeants finished talking with me, I had no doubt about what I needed to do next. I flashed back to what I learned from my Ranger commanders when I was a lieutenant about the importance of a healthy leadership climate and how it is the responsibility of all leaders to take action when they detect or are told about an unhealthy climate.

> *"A leadership team has to work together by always sharing what they see." He then told me to think of myself as one of his climate monitoring stations. "If you see a tornado touch down, you have a responsibility to share that information so the rest of the leadership team can take action and make sure none of our people get hurt by it. By not reporting the tornado, you're allowing it to destroy the next town in its path and the next one after that."*

"I'll never allow it to happen again, Sir, you have my word," I told my battalion commander at the time. Now I just had to find the right person to talk to.

The next night (which was my last at Baghdad International Airport), I was standing outside our tactical operations center talking with a couple of my guys who had just returned from a successful capture operation, when I spotted a senior Ranger officer, who I believed was one of only two or three officers on the ground in Iraq at that time who had both the command authority and the rank authority to investigate and do something about whatever was going on in the 2nd Ranger Battalion. I figured it was as good a chance as I might get since my guys and I were scheduled to move to our new safe house a few hours after sunrise.

It's always a dicey endeavor trying to tell a fellow leader that one of their subordinates, or one of their subordinate units, has potential leadership issues and/or bad morale. In situations such as these, a leader's ego is usually their biggest enemy. No matter how self-actualized you may think you are as a leader, it's a tough pill to swallow when someone walks up to you out of the blue and tells you that everything you thought you knew about a subordinate unit or subordinate leader is not as it appears to be.

Yet whenever I look at the situation through the lens of common sense and ask myself whether I would want someone to tell me about possible leadership or morale issues under my command, I always come up with the same answer. **Of course I would want the person to tell me everything they know, good, bad, or indifferent. If there is any chance it's true, then it means my people are crying out for help, and it's my responsibility as their leader to look into it and take appropriate action based on what I learn.**

Once again, I recalled the words of one of my former Ranger battalion commanders: "As a leader, you can be both tactically and technically competent, and a super nice guy who is fun to be around, but if you turn a blind eye to a tyrannical or incompetent subordinate who is making people's lives miserable and preventing them from accomplishing their purpose, then you are an enabler, and your people will despise you with the same vigor they despise the tyrant."

After sharing some late-breaking operational updates, I explained to the senior Ranger officer that "I had been talking with a couple of senior sergeants from the 2nd Ranger Battalion, whom I'd known and respected for many years, and they shared some pretty significant things about the leadership climate in the battalion that I thought you might want to hear." "OK," he responded.

As quickly and succinctly as possible, I recounted what Sergeant X and the two other NCOs had told me. I chose my words carefully for two reasons: I wanted to ensure that what I was telling him matched as closely as possible with what Sergeant X and the others told me, and I also wanted to do it in a way that didn't gloss over the seriousness of the downstream effects it could cause.

I reinforced that I didn't know any of the current leaders in the 2nd Ranger Battalion, so I had no idea if the problem was limited to one single individual or to a group of individuals. I added that "it sounded like there was at a minimum some serious morale issues in the battalion and, based on the credibility of the three NCOs who shared the information with me, some serious readiness issues that could affect the battalion's ability to operate effectively in combat."

It was a few hours before sunrise and still dark outside, so I couldn't see his face as I spoke. His silence and lack of feedback told me I should stop talking and listen to what he had to say. He leaned in close, tapped his finger on my chest, and repeated two key things I had just said: "You aren't in the Ranger regiment anymore," and "you don't even know any of the members of the chain of command they're talking about." After a few long, emotional hyperlinks off both points, he ended the conversation with the words: "If I had a nickel for every time a disgruntled sergeant complained to me about following orders and doing what they are told to do by their commanders I'd be rich, so why don't you tell your anonymous sources to stay in their lanes and do what they're told to do by their chain of command." And with those final words, he bid me farewell and walked away.

I can't say I was shocked by his response because I didn't have any specific expectations regarding how he'd react other than that he'd listen, consider the context, and then go check it out. It seemed then, as it does now, like common sense. Despite his emotion-based reaction, I still believed that with a bit of reflective time and experience-based feedback, he would come to his senses and check it out. **After all, even the most self-centered or selfish of leaders should eventually recognize that a toxic leadership climate has no possible upside for them and huge potential downsides in the areas of morale and welfare of their people, as well as the unit's capacity to accomplish any type of real-world mission.**

A few days later, and just over a month after they had arrived in Iraq, the 2nd Ranger Battalion abruptly redeployed back to their home base at Fort Lewis, Washington. I never saw or talked to the senior Ranger officer again, nor did I have another chance to talk face-to-face with Sergeant X. Our request for Ranger augmentation had been denied by higher headquarters.

The best way for me to describe the next six weeks was as a combat blur. We averaged over three capture missions per night, and most of our targets were located in urban areas, which meant lots of places for the enemy to hide, lots of private property, and far more interactions with innocent civilians than bad guys captured or killed. There was plenty of work to be done and nowhere near enough operational personnel on the ground to do it.

Despite our shortage of manpower and obvious need for Ranger augmentation, I felt relief when I heard that the Rangers had been sent home. I assumed the reason for their abrupt departure had to be related to the leadership climate issue.

Six weeks later, I rotated out of Iraq and spent the next eleven months working as a part of the Secretary of Defense's Corporate Fellowship Program. Established in 1981, the original purpose of the program was to exchange leadership and management best practices between military and corporate leaders by sending eight military officers (two from each service) to work for one year on location at some of our country's most successful corporations. I was sent to a biotechnology company in Southern California, where, for all practical purposes, I was "off the military information grid."

As a result, I had no idea what, if anything, the senior Ranger officer I had talked to that night did or didn't do with regards to following up and/or taking action on the leadership climate issue. *Maybe he looked into it and didn't think it warranted further action. Maybe he looked into it and worked with the leadership team to make things better. And, of course, maybe he did nothing and the Rangers did what Sergeant X told me they always do: "sucked it up" until they got an entirely new chain of command.*

Again, my rationale for anonymizing his name and sharing details of the above conversation are the same as for everyone else involved in the sequence of events that follows. The purpose of these pages is to neither judge, adjudicate, nor to infer blame. The purpose is to learn from it. The human brain learns from experience. If you haven't experienced a similar situation (directly or indirectly), you can't have learned from it. Perhaps someone reading this right now will face a similar situation in the future and remember that they have **a responsibility to take action by sharing what they learned with their fellow leaders, or bear the potentially tragic ramifications of inaction.**

When the Rangers redeployed from Iraq, I felt like I had done everything I could have done given the circumstances. One year later, while sitting in my office at the Pentagon and staring at the flash report headline, I didn't have any reason to believe there was a relationship between the leadership climate in Iraq and what had just happened to Pat Tillman in Afghanistan. I never considered that the same Ranger chain of command that was responsible for the leadership climate in Baghdad might still be in place one year later in Afghanistan. **In retrospect, I also never comprehended that a toxic leadership climate could directly or indirectly cause accidents, injuries, and even death.**

I didn't hear anything else regarding the circumstances surrounding the death of Pat Tillman in those first few weeks. My boss at the Pentagon was my good friend and fellow common sense sentinel, Tom, and I'm pretty certain it wasn't a matter of access to information because I had access to whatever was coming in.[26]

I was also actively searching.

To put that period of time in context, there were 254 Americans killed in action in Iraq from January through April of 2004 and an additional 595 killed from May through December of that year.

Another fifty Americans were killed in action in Afghanistan during the same period. Over a thousand more were wounded. This made 2004 the highest casualty-producing year for the American military since the beginning of the Global War against Terrorists in 2001.[27]

The complexity of commanding and controlling military forces fighting in two different countries against two different enemies and the political pressures that went along with it were unprecedented. I had never worked in the Pentagon before, so I didn't have any experience to compare it to. I just remember how chaotic it was and how difficult it was to get the attention and/or focus of high-level leaders (military or civilian) on any specific issue for more than a few minutes or hours.

The first official announcement regarding the incident occurred on 29 May (almost five weeks to the day after it happened), when the Army officially acknowledged, "friendly fire probably killed Pat Tillman."[28] The story received headline coverage across all media outlets. Instead of being surprised by the friendly fire announcement and the stories that followed, I was surprised that so many others were.

My surprise was based on two faulty assumptions: I assumed that everyone had the same "sounds like friendly fire" reaction as I did to the initial reports of Pat Tillman's death. I also incorrectly assumed that the Army had kept the Tillman family abreast of the facts as they learned and discovered them. That they hadn't was more than surprising; it was deeply disturbing and monumentally embarrassing.

What did the Army tell the Tillman family about the circumstances of Pat Tillman's death during those first five weeks? According to Pat Tillman's brother Kevin, who was a member of the same Ranger platoon as Pat but was separated from him when he was killed: "It started at the May 3rd memorial service when the Army released a manufactured Silver Star narrative detailing how

Pat died while charging up a hill and leading a courageous counterattack in an Afghan mountain pass." Mary Tillman told The Washington Post, "The fact that he was the ultimate team player and he watched his own men kill him is absolutely heartbreaking and tragic. The fact that they lied about it afterward is disgusting."

Tillman's father, Patrick Tillman, Sr., was incensed by the false narrative surrounding the cause of his son's death, which he attributed to a conscious decision by the leadership of the U.S. Army to protect the Army's image: "After it happened, all the people in positions of authority went out of their way to script this. They purposely interfered with the investigation; they covered it up by presenting outright lies to the family and to the public."

Why would the Army cover up the details surrounding Pat Tillman's death from his family? I asked myself. It didn't make sense to me or to any of my colleagues in Special Operations that I talked to about it. The only known details about what actually happened came from the first official investigation that was completely compartmented within the Ranger regiment and the Joint Special Operations Command one level above them. The only rationale I could come up with for why the military might hide the details surrounding Pat Tillman's death was if the details revealed sensitive intelligence pertaining to ongoing operations.

In the months that followed (June 2004 through August 2005), I traveled to both Iraq and Afghanistan as a member of a security assessment team and spent extensive time on the ground in both countries. It was during this period that I learned that the mission the Rangers were on when Pat Tillman was killed had nothing to do with the hunt for Osama bin Laden, nor was it related to any of Osama bin Laden's underlings. When I asked a friend of mine who was serving as a staff officer in the Rangers' higher headquarters, "What were the Rangers actually doing when it happened?" He told me, "They were just driving around looking for weapons caches." He also informed me that the Army had launched a second investigation of the incident, which was ongoing at that time.

The Tillman family didn't want another investigation from the Army—they simply wanted truthful answers. **Specifically, if Pat Tillman was killed by friendly fire and the Army knew it was friendly fire within hours of his death, why didn't they tell Pat's brother Kevin and the rest of the family what they knew?** As Mary Tillman continually asked, "Why in the world did they make up such an elaborate story?"

Fed up with the lack of answers she was getting from the military, Mary asked her congressman and senator for assistance. Within weeks, they directed the Army to send declassified copies of all investigations and interviews to the Tillman family so they could see the results for themselves. Mary dove into the documents with the verve of a detective and the veracity of a mother who's been lied to about the loss of her loving son. Although the documents were heavily redacted and not organized chronologically, she combed through every word on every page. Then she tabbed and highlighted them. Once it was possible to make sense of the details, she began sharing copies with select journalists in hopes that their involvement would create awareness and assist her family in learning the truth about what happened.

In December of 2004, an article was published by author Steve Coll that provided a comprehensive account of the sequence of events leading up to, during, and after the firefight that killed Pat Tillman.[29] The article included maps of the terrain, icons that showed the different vehicles involved, and the sequence of events arranged in chronological order. The article also included something that rarely sees the light of historical scrutiny: transcripts of conversations between Pat Tillman's platoon leader and his higher headquarters during the moments leading up to the firefight. In retrospect, the conversations provided the first hint that this wasn't just a one-off case of battlefield friction or the fog of war. Even though the conversation took place over modern-day email, the dialogue was painfully familiar to veterans of all wars and a prime example of the senselessness of hierarchical decision-making and micromanagement in combat.

In December of 2006, Mike Fish of ESPN published another comprehensive and thoroughly researched account of what happened. This article included taped conversations with some of the Rangers from the platoon as well as video reconstructions that revealed the time/distance relationship between the different elements of the platoon that were involved in the firefight.

My first reaction after reading these articles and viewing the pictures and videos was a question: **Why do members of the military, who are risking their lives every day, have to depend on third-hand accounts of key events to learn lessons with such obvious life-saving relevance?** There was a ready-made and rapt audience for these lessons as vehicle-mounted firefights and ambushes were occurring with increasing frequency in both Afghanistan and Iraq. Critically, some of the deadliest vehicular ambushes and firefights were still ahead of us.

In 2009, Jon Krakauer published a book on the incident titled Where Men Win Glory. In it, Krakauer renders a well-constructed account of the events and actions that led up to Pat Tillman's death, using information from the four investigations as well as passages from Pat Tillman's personal journals, interviews with his wife and friends, and conversations with the soldiers who served alongside him.

Although Krakauer's book came out in 2009, I didn't read it until a friend gifted it to me in 2017. As I began reading, I still didn't have any reason to believe there was a connection between my conversation with Sergeant X regarding the leadership climate at Baghdad International Airport in April of 2003 and what happened to Pat Tillman one year later in Afghanistan in April of

2004. Three passages from Pat Tillman's journal that were published in Krakauer's book made the connection for me.

Journal entry #1 dated 30 April 2003:

> *"Four Delta Force operators were shot while on a mission to capture a 'high-value target,' and I helped carry one of the wounded soldiers in from the medevac helicopter to receive treatment. The man I was carrying had been shot in the abdomen. At this point in the game, I was quite surprised to see anyone shot... The danger seemed minimal. It goes to show you never know."*

Journal entry #2, written after the battalion conducted a mission in Baghdad, dated early May 2003:

> *"We've had leaders telling guys to shoot innocent people only to be ignored by privates with cooler heads....it seems their battlefield sense is less than ideal. Given the stress of a situation, I absolutely will listen to my instincts before diving headfirst into any half-baked scheme of theirs. Perhaps this is not the 'military right,' however, these past couple of months have suggested it's necessary."*

Journal entry #3, dated 14 May 2003:

> *"We are leaving at 0300 tomorrow, Thank Fucking God."*

"Holy shit, he was at Baghdad International Airport," I said out loud to myself as I read his journal passages. I would later learn that the other Rangers who helped carry stretchers with Pat Tillman were his teammates from the 2nd Platoon.

In his book, Krakauer attributes the three journal entries to Pat Tillman's philosophic position on the senselessness of the war in Iraq. Neither I nor anyone else can credibly comment on what Pat Tillman was actually thinking when he wrote those passages in his personal journal. Yet given what Sergeant X and the other two senior sergeants told me, it seemed far more likely to me that his journal comments were related to the 2nd Ranger Battalion's leadership climate than to his philosophical position on the war and why we were there.

As previously mentioned, I have kept a pocket notebook with me for most of my military career. I never call it a journal; instead, I refer to what I write as, "Thoughts, Insights, and Lessons." When I go back and review my notes, I find that some of them are amazingly insightful, some were driven by my first impressions and/or by incomplete information, and some were written in a way that didn't accurately reflect what I actually learned. The point is that the only one who can interpret personal journal notes is the person who wrote them. As such, I didn't draw any conclusions or attempt to interpret Pat Tillman's other journal entries.

The key takeaway for me was that he was at Baghdad International Airport in 2003 at the same time Sergeant X told me about the leadership climate issues in the 2nd Ranger Battalion. In addition to the three journal entries, Krakauer's book also contained the names and ranks of the entire Ranger chain of command, which further underscored the connection. I finally understood that key members of the Ranger chain of command that were in place at Baghdad International Airport in 2003 were also in place in Afghanistan at the time of Pat's death in April 2004.

After I finished reading Krakauer's book in 2017, I reflected on the fact that the military had spent millions of dollars and thousands of man-hours to conduct and complete four separate investigations into the tragedy. Yet when I asked myself what, if anything, we learned from it all, I couldn't come up with a single example. **Perhaps the second biggest tragedy of the Pat Tillman**

incident was this. That thirteen years on, we didn't seem to have learned a thing from it.

It was during this time that I began discussing what I had discovered with some former Rangers who I was fortunate enough to stay in contact with and, more importantly, still call friends. All three of them had served in the 2nd Ranger Battalion during the period between 2003-2004. The more I talked with them, the more context-specific knowledge I learned.

One of my friends, Sergeant Major (retired) Ted Kennedy, who spent most of his career in A Company 2/75 and with whom I had served during both of my tours, told me that despite all the above-mentioned articles and books, there was still a lot of information that hadn't come out in the investigations. He also told me that Mary Tillman wasn't satisfied "that she knew the truth about what actually happened to her son," so she was still doing research and collecting information in an effort to put all the pieces together.

When I asked Sergeant Major (ret.) Kennedy, who was serving as first sergeant of A company while they were at Baghdad International Airport, about the leadership climate during that time-frame, here's what he said: "Unfortunately, I can remember it like it was yesterday; everyone in the battalion was frustrated." The brain doesn't remember pain, but it does remember frustration. "With regards to the conversation you had with the senior Ranger officer about looking into it. I would have hoped a professional would have handled it another way, but some leaders can't stray far away from the culture of alphas…finger poking and all, for better or worse!! Many of them have very small toolboxes indeed, and some have brains to match!! Just intimidation by rank, the big-hammer bullies like to use. My opinion."

He concluded by telling me: "My last three years in the Rangers contributed greatly to making me the man I am today, RETIRED! It also gave me the mantra, of 'if I'm not happy now, it's my fault.'

Lots of stories from that place…and lots of walks to cool off to make sure I didn't make decisions that furthered the badness and caused the soldiers more misery. Sad part is, the leaders could not see it themselves, as they spent most of their time pointing out all the failures of others. Even though I'm out of the Army now, I constantly think about young men and women who, like the Tillman brothers, volunteer to risk their lives in service to their country and end up having to spend their time in the military struggling under toxic leaders. I think the problem of toxic leaders and toxic leadership climates is far worse today than most of us realize."

My friend's wise words about toxic leaders and the toxic leadership climates they create made me realize I didn't know enough about what toxic leadership actually is, or how pervasive it is across all leadership climates. It also made me realize that common sense leaders have a responsibility to our country, our companies, and our people to understand the problem and figure out a way to fix it instead of burying our heads in the sand and hoping it will go away.

CHAPTER 5

What is Toxic Leadership?

*Experience and time reinforce that leadership
is learned. Every leader deserves a chance
to learn how to be a better leader.*

To solve any type of complex problem, we have to ask ourselves the right questions first: What is toxic leadership? Is toxic leadership really a bigger problem than most of us ever thought? How do we recognize the symptoms of toxic leadership when we see them? How can we prevent new leaders from becoming toxic? And what can we do to prevent toxic leaders from poisoning the people, the places, and the purpose for which they are given the privilege to lead?

What is toxic leadership? In 2003, after a pattern of high-profile incidents involving toxic leaders during combat operations in Iraq and Afghanistan, Secretary of the Army Thomas White directed the Army War College[30] to look into ways the Army could effectively assess leaders and detect those who might have destructive leadership styles.

The Army War College immediately began studying recent real-world examples of toxic leadership while also surveying 172 officers from all four branches of the military on the topic.

Colonel George Reed was a student at the Army War College who participated in the study and then went on to write and publish a groundbreaking paper on the topic, appropriately titled "Toxic Leadership."[31]

When I talked with George Reed, I told him how impactful his paper was to me and asked him why it was never shared with the rest of the Army in 2003. He explained, with a hint of frustration, that, "Shortly after we published the paper, the Secretary of the Army Thomas White retired, and unfortunately, the priority to take action on toxic leadership retired with him." I thanked him for his work and told him, "My goal is to bring your paper and the priority to eradicate toxic leadership out of retirement."

What is a toxic leader? The phrase "toxic leader" is linked with a number of dysfunctional leadership styles.[32] Today, the terms "toxic leader," "toxic leadership climate," "toxic culture," and "toxic organization" appear with increasing frequency in both military and business leadership literature. **It's important to note that there can be more than one toxic leader who contributes to the creation of a toxic leadership climate and culture.** Other leadership styles often used synonymously with "toxic" include: tyrant, bully, A-hole, prick, and boss-from-hell.

Dr. Jean Lipman-Blumen, Professor of Organizational Behavior at Claremont Graduate University, explains that toxic leadership is not about run-of-the-mill mismanagement or incompetence. Rather, it refers to leaders who, by virtue of the way they think, decide, and behave, "inflict reasonably serious and enduring harm," not only on their own followers and organizations but also on others outside of their immediate circle of victims and subordinates. This definition underscores that there is no upside to toxic leadership. Toxic leaders have poisonous effects on the people, the place, and the purpose for which they've been given the privilege to lead.

What is the scope of the problem? The prevalence of toxic leaders and toxic leadership climates in the military is difficult to assess. As George Reed points out: "survey data on toxic leaders are biased because they underreport the actual numbers as only the 'survivors' are polled." Unfortunately, toxic leadership climates are all-too-familiar amongst those survivors. Virtually every Army War College student participating in the project had spent time serving under at least one toxic leader. What follows is a sampling of feedback from some of those experiences:

> *"It never ceases to amaze me how they are able to fool everyone above them when everyone below them knows they should be thrown out of the military."*
>
> *"All you have to do is think of the men and women that have to live their lives every day under these leaders to realize that we have a responsibility to our country and our soldiers to stand up for them and take action instead of burying our heads in the sand."*
>
> *"We have got to do something as a moral imperative to identify these leaders and stop them from going forward. The higher they are in the system, the more damage they do, and the more they can spread their toxicity."*

To suffering subordinates, toxic leaders represent a daily challenge that can result in unnecessary organizational stress, negative values, and hopelessness. As Reed explains, "toxic leaders do not add value to the organizations they lead. They have a negative impact on unit climate, culture, and *esprit de corps*. At one end of the spectrum, toxic leadership climates erode trust, reduce effectiveness, and affect soldier well-being and retention; at the other end, under the most extreme conditions, **toxic leadership climates can lead to mutiny and even death**."[33]

*"I'm telling you all this because the leadership climate is
so toxic and everyone is so stressed out about it that I'm
concerned something really bad is going to happen."*

In her analysis of leader toxicity in corporate environments,
author Kathie Pelletier identified other potential consequences,
including: workplace deviance by subordinates, **increases in
churn and turnover rates, and chronic psychological stress,
which leads to deteriorations in performance and morale.**[34]

A more recent study conducted by the Center for Army
Leadership[35] which surveyed over 30,000 military and civilian
leaders, found that one in five leaders is seen as toxic by their subordinates, and a staggering 83% of those surveyed reported they
had interacted (directly or indirectly) with toxic leaders in the last
year. The survey data also revealed that toxic leadership is not just
an Army issue but a problem that is pervasive in all four military
services. Only 11% felt that the problem of toxic leadership is not
severe.

How do we recognize toxic leadership? Real-world examples[36] of military commanders relieved during combat operations
in Afghanistan and Iraq reveal that there is no common demo-
graphic or behavioral pattern shared by toxic leaders. While this
may surprise some members of the military, it's not surprising
to those who have spent significant time in the corporate world,
where toxic leaders are just as common and come in all sizes,
shapes, colors, genders, nationalities, and preferences.

According to Reed, "Leader behaviors such as loud, overbearing, and demanding are not specifically correlated with being toxic.
A soft-spoken leader with calm demeanor and a façade of sincerity can also be toxic." Most researchers agree that **there is not one
specific pattern of behavior that deems a leader toxic; instead,
it's a combination of dysfunctional behaviors manifested by the**

choices they make and how those choices affect the people they lead. According to the 30,000 survey respondents, what follows are some of the most common patterns of behaviors seen in toxic leaders. The list is not all-inclusive.

1. **Suffocating Micromanagement (aka Control Freaks): What is it?** The concept of micromanagement can be extended to any social context where one person takes an overbearing or bullying approach in the level of control and influence they exert over other members of a group.

   What are the Symptoms?

   - A micromanager tends to require constant and detailed performance feedback such as reports, updates, and PowerPoint presentations that focus excessively on procedural trivia, often in greater detail than the micromanager is capable of understanding. Micromanagers aren't able to see the forest because they are obsessed with counting the leaves.

   - Constant meetings and inspections are another common symptom of micromanagement that is often overlooked. When leaders require numerous time-consuming meetings, inspections, and video teleconferences in order to "give guidance," "get warm and fuzzy," and/or "get approval," **they are indirectly telling subordinates that they don't trust their abilities or their decisions**. The organization pays a heavy price for their toxic leader's obsession with managing minutiae in the form of opportunity costs.

   - Micromanaging is oppressive, fosters anxiety, and creates a high-stress work environment. Eventually, employees will become disenchanted and quit to work for another organization.

2. Mean-Spiritedness/Sociopath:

What is it? In her book, *The* Sociopath Next Door, Dr. Martha Stout of the Harvard Medical School declares that four percent of the world's population are sociopaths. That's one out of every twenty-five people worldwide who have no conscience, no sense of right or wrong, and no empathy, yet possess a dangerous ability to mimic emotion and empathy as a means of ingratiating themselves and manipulating others for their benefit or amusement.

What are the Symptoms?

- Fails to acknowledge responsibility and always deflects blame onto others.

- No real feelings, empathy, or compassion exist. After sabotaging, firing, or ruining a person's career, they have no remorse. In extreme situations, they may celebrate the situation with comments such as, "they had it coming," or "that's what happens when you disobey/screw up," or "that will teach them."

- They know right from wrong but couldn't care less. They manipulate and bully people with the decisions they make and actions they take. To the sociopath, what's right is whatever is in their own best interest. If it serves them well, they believe their actions are completely justified.

- Their worst tendencies quickly surface when they feel they are losing control over you. When a sociopath finds out that you are not on board with or supportive of them, they'll move into attack mode to make your life miserable while criticizing you to anyone who will listen.

3. Lack of Respect/Trust:

What are the Symptoms?

- It starts with language and common courtesies such as the use of "thank you," "please," and "sorry about that" when conversing with subordinates. When common courtesy is present, it demonstrates care and concern. In many toxic climates, these courtesies are considered unneeded "fluff" and a waste of time.

- The lower you are in rank, seniority, or title, the less respect there is for you. In these environments, soldiers, non-commissioned officers, and junior officers are constantly reprimanded for not accomplishing tasks in a way that meets their toxic leader's always changing standards. Respectful leaders talk and act the same way around privates as they do around generals. Disrespectful leaders act like two different people.

- Rudeness is more common than civility. Studies show that people exposed to rude behavior are more likely to interpret random but benign future behaviors as rude. Critically, they are more likely to behave rudely toward other people, which creates hostility and makes those people more likely to act rudely to the people they come in contact with afterwards. Rudeness is contagious.[37]

4. Lack of Communication:

What are the Symptoms?

- Disrupted information flow. There is no learning-feedback loop. Orders, emails, and phone conversations are all one-way. Purpose and priorities are "often ambiguous," and when they are present, they rarely make sense. In a toxic leadership climate, knowledge that goes up rarely, if ever, comes back down.

- No sense of urgency for information dissemination. Information is withheld until the last possible moment, with little concern as to how it will affect subordinates. "Heads up" or "Warning" orders to initiate activities and movements aren't sent out until the last minute.

- No direct communication with subordinates. Everything goes through the hierarchy, staff, or electronic messaging. Orders, commands, and directives are issued without the logic of why they make sense. "Because I said so," "just do what you're told," and "you don't have a need to know" are common refrains from toxic leaders when a subordinate asks why. The inability to speak up, out, or with leaders to ask clarifying questions leads to confusion amongst subordinates, continuous uncertainty, and stress.

How do leaders become toxic? Doctor Karen Wilson-Starks, an expert in corporate leadership, suggests that toxic leaders emerge because they themselves were mentored by toxic leaders. As Reed points out, "If left unchecked, toxic leaders create a self-perpetuating cycle of toxicity with harmful and long-lasting effects on morale, productivity, and retention of quality personnel." **Toxic leadership is learned.**

A well-known pattern of abusive behavior is that it's contagious. Studies indicate that between one-third and one-half of people who experience abuse in childhood will become abusers themselves. One year under a toxic leadership climate is more than enough time to create a culture where subordinate leaders up and down the chain become conditioned to a way of thinking and making decisions that will appease their toxic boss instead of taking care of their people. Along the way, subordinates also learn to mimic their boss' toxic behaviors, especially when the leader is present. **Toxic leadership is contagious.**

Why don't toxic leaders get called out? There are many reasons toxic leaders are able to survive without getting called out. What follows are three of the most common:[38]

1. **Culture and Values:** Military culture and values emphasize automatic respect for rank and title, even if the leader has done nothing to earn that respect. Every soldier who joins the military does so by raising their right hand and saying out loud the Oath of Enlistment:

   > *"I (state full name) so solemnly swear (or affirm)*
   > *that I will support and defend the Constitution of*
   > *the United States against all enemies, foreign and*
   > *domestic; that I will bear true faith and allegiance to the*
   > *same; and that I will obey the orders of the President*
   > *of the United States and the orders of the officers*
   > *appointed over me, according to regulations and the*
   > *Uniform Code of Military Justice. So help me God."*

   Because of the deeply ingrained sense of loyalty to their unit and the leaders within, soldiers are often conflicted about speaking up or telling their boss' boss about toxic leaders. The fact that the toxic leader is almost always a key contributor to the suffering subordinate's end-of-year rating/review makes speaking up about their toxicity a potentially career-ending endeavor.

   If the suffering subordinate does speak up or out about a toxic leader and nothing of substance comes of it, the subordinate risks losing the trust and loyalty of the leaders above and around them.

2. **Administrative and Personnel Policies:** Administrative policies that create fixed patterns of personnel movements, such as the U.S. Military's policy of changing out

unit commanders every two years, perpetuate a "let's wait it out; it's only a matter of time before they leave" mentality amongst subordinates. Personnel rating policies that emphasize and reward short-term administrative results are also a factor because short-term administrative tasks are toxic leaders' specialty. Policies designed to focus on harassment and proportionality protocols also enable toxic leaders. These can be used as a shield to divert attention and protect toxic leaders from actually being called out as such.

3. **Apathetic or Self-Centered Leaders Above Them:** In many cases, toxic leaders are enabled by the leaders above them, who may or may not be toxic themselves. Why would a non-toxic leader turn a blind eye to a toxic leader? Although the rationales are many and varied (lazy, incompetent, naïve, etc.), they all have one thing in common, toxic leaders make their bosses look better and/or their lives easier.

Toxic leaders are experts at ingratiating themselves with their bosses. And because they lack compassion and empathy, toxic leaders let their bosses know they can be called on at any time, day or night, to do menial administrative tasks such as, inspections, presentations, and reports; or controversial personnel-oriented tasks such as: reprimanding people, reducing agreed upon compensation, or demoting and firing people. Self-centered leaders turn a blind eye toward their toxic subordinates' behaviors because the toxic subordinate will do these types of tasks whenever needed, without questions, without remorse, and without requesting the logic of why they do or don't make sense.

Conclusion: The military must make eradicating toxic leadership from its ranks a tangible priority by changing its current culture of toxic tolerance to one marked by toxic intolerance. Policies and procedures are a critical starting point; however, it's the leaders, at every level, who have the power to change the culture by embedding an anti-toxic ethos in every choice they make and every action they take.[39]

While 83% of the respondents of the Center for Army Leadership survey said they had directly observed a toxic leader during the prior year, 97% had also observed an exceptional leader. This reinforces that there are more than enough highly competent, highly motivated, common-sense leaders to eradicate the toxic leaders who rule by fear, intimidation, and coercion.

Superiors are in particularly important positions to deal with toxic behavior, as they are the ones with the authority to counter it. Yet they may also be the last to observe the behavior unless they are taught to tune into it. Although subordinates are generally not in a position to fix the problem, they are the first line of defense, and, as such, must be made aware of how to recognize and report the problem. Critically, they need to hear from the highest-ranking leaders in their chain of command that toxic leaders and toxic leadership climates are not to be tolerated.

In both military and corporate environments, toxic leadership climates are more than just a nuisance; they are the equivalent of having a saboteur within your ranks. Toxic leaders poison the people, the place, and the accomplishment of the purpose for which they were given the privilege to lead. Not having toxic leaders within your system is a proven competitive advantage, as is going up against a competitor that has toxic leaders within theirs.

After completing my research on toxic leadership, I had a better understanding of how significant the problem is and how serious the effects can be on an organization, its people, and their ability to successfully accomplish any type of mission. It was at this point that I realized I needed to share what I had learned with Mary Tillman.

CHAPTER 6

How a Toxic Leadership Climate Leads the Platoon Into Chaos

*Complex systems have what is known as sensitivity to ini-
tial conditions. The classical formulation of this comes from
Edward Lorenz, a meteorologist who was one of the first, in
1963, to investigate the properties of complex systems such
as weather. Now known as **Chaos Theory**, it states that even
such a small perturbation as a butterfly flapping its wings
in Brazil could—because of the nonlinear nature of the system—lead to a tornado some months or years later in Texas.*

7–9 April 2004: 2nd Ranger Battalion deploys from Fort
Lewis, located near Tacoma, Washington, to Bagram Air Base in
Afghanistan. (See map.)

12 April, 0200 A.M.: Alpha Company with approximately 150
men flies via four CH-47[40] helicopters from Bagram Airfield to
their forward operating base (FOB) in Khowst.

13 April: After landing, the forty-four members of 2nd
Platoon immediately began loading weapons, ammo, water, and
equipment onto their vehicles and conducting communications

Map of Afghanistan

checks on all their radios. Within a few hours after landing, the 2nd Platoon was driving out the front gate of their forward operating base in Khowst. Their destination was Spera District, located forty-four miles to the southwest. (For more information on how a Ranger platoon is organized, see annex A: Military Rank and Organization Annex)

What was the Rangers' mission or purpose? Most official accounts state that the Rangers' mission was to "kill or capture anti-coalition forces within their area of operations." However, kill or capture is not an accurate description of their actual purpose. Kill or capture operations are focused on killing or capturing a specific individual or individuals. The mission of the 2nd Platoon was far more ambiguous. As part of a larger operation known as Operation Mountain Storm, the 2nd Platoon's mission was to:

Divide the region along the Pakistan border into numbered grid zones (e.g., twenty-five through fifty), then drive to the villages within those grid zones and conduct house-to-house searches for insurgents and weapons caches. Once they

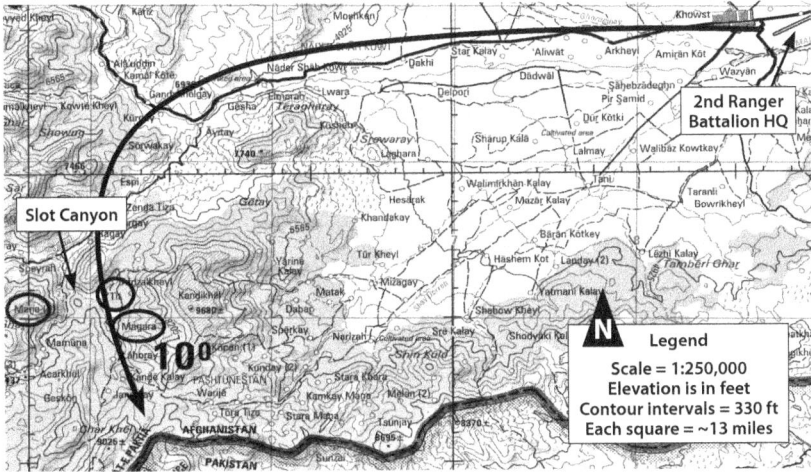

Map of 2d Platoons route from their forward operating base in Khowst, to their area of operations in Spera district. Note proximity to the Pakistani border

searched a village and confirmed there were no insurgents or weapons caches, they would pronounce the village as clear and drive to the next grid zone to do the same thing. Progress was monitored and measured by how many grid zones the platoon could search and call in as clear (e.g., "grid zone twenty-five secure, moving on to grid zone twenty-six").

Information regarding the overall purpose of Operation Mountain Storm was either redacted or not included in all four official investigations. However, while doing research to better understand why these "airborne" Rangers were conducting a "motorized" Ranger mission in one of the most un-trafficable and inhospitable areas in Afghanistan, I came across an open-source article published by Asia Times Online. The article is based on interviews with Afghan resistance fighters who provide context-specific knowledge of the Operation Mountain Storm plan, as well as the enemy, the terrain, the civilian populace, and the history of the area the Rangers would be operating in.

ASIA TIMES ONLINE LTD.

"Grand Plans Hit Rugged Reality"
By Syed Saleem Shahzad

20 March 2004

The plan to eradicate the Afghan resistance was straight-forward: U.S.-led coalition forces would drive from inside Afghanistan into the last real sanctuary of the insurgents, and meet the Pakistani military driving from the opposite direction. There would then be no safe place left to hide for the Taliban and al-Qaeda remnants, or, presumably, for Osama bin Laden himself. The plan's implementation began with the launch of operation "Mountain Storm" around March 15.

But the insurgents have a plan of their own, which they have revealed to Asia Times Online. Conceived by foreign resistance fighters of Bangladeshi, Pakistani and Arab origin, it is a classic guerrilla stratagem that involves enmeshing the mighty military forces of the United States and its allies in numerous local conflicts, diverting them from their real goal and dissipating their strength.

In an exclusive meeting with Asia Times Online, a prominent planner of the Afghan resistance spelled out the strategy. Pointing to a hand-drawn map, the insurgent indicated an area he called "Shawal." Technically speaking, "Shawal" falls on the Afghan side of the Durand Line that divides Pakistan and Afghanistan. (Editor's note: The border area inside North Waziristan is also called Shawal.) In reality, "Shawal" is a no-man's land, a place no one would want to go to unless he were as tough as the local tribespeople, a guerrilla fighter taking on the U.S., or, perhaps, Osama bin Laden. Shawal is a deep and most dangerous maze. The insurgent described it thus:

"One crosses the first mountain and sees a similar mountain emerge and after crossing another mountain he feels a spin in his head and thinks the whole world in this area is the same and leads the way nowhere."

This is the last safe haven for the Afghan resistance, from which they launch attacks on coalition forces and the Afghan government, and to which they return to regroup and receive sustenance from the locals. And this is the kind of terrain the U.S. and its allies will encounter in their drive to occupy "Shawal" whether they come in from the Afghan side via Bermal, or from the Pakistan side via South or North Waziristan.

Those who are masters of this maze can raid the Afghan provinces of Ghazni, Paktia, and Paktika. The only masters are people of the Data Khail and Zaka Khail tribes and the insurgents who base themselves there.

The Data Khail and Zaka Khail have a long history of defiance and have never capitulated to any intruder. The tribesmen are as tough as the terrain, and they have been known for centuries for their strong bonds of loyalty, such that "even an enemy who requests shelter would be given it." These two tribes are now the protectors of the Taliban and al-Qaeda fighters based in "Shawal." **By occupying the area, the U.S. hopes to deprive the insurgents of the tribes' crucial support. Forced to flee, the insurgents would eventually fall into the hands of the United States' local proxy networks of anti-Taliban tribes and warlords. Such is the plan."**

As noted, this article was posted online on 20 March 2004, three weeks before the Rangers drove through the gates of their forward operating base in Khowst. The area the insurgents describe in the article as "Shawal" is the same area the Rangers were tasked with clearing. Which means that Operation Mountain Storm and the Rangers' mission was compromised and known to the enemy before it even began. None of the Rangers I talked to had ever heard about or seen the Asia Times article. There are no references to the article in any of the investigations, which leads me to believe that the U.S. military was unaware of its existence both before Operation Mountain Storm began and after it ended.

When they drove out the gate of the forward operating base in Khowst, the Rangers of 2nd Platoon were riding in a convoy consisting of nine vehicles: four gun-mounted Humvees (GMV), one command Humvee, two cargo Humvees, and two Toyota Hilux SUVs (See vehicle schematic.)

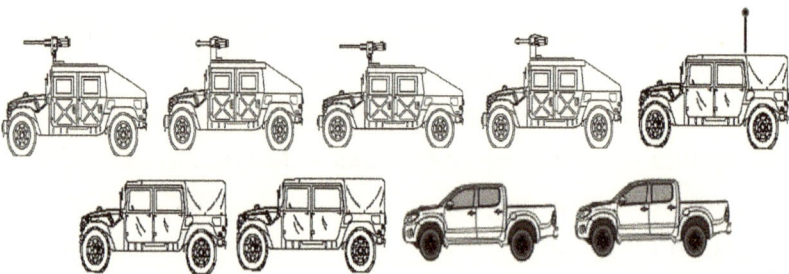

Corporal Tillman (seated on the right) and his squad riding in their Gun-mounted Humvee, while "clearing zones" near the Pakistani border. The squad leader (standing) rides in the front passenger seat. The Ranger private from Corporal Tillman's team is seated on the left side. The weapon mounted in the middle is a MK-19. A few days after this photo was taken, the vehicle would breakdown and the platoon would be ordered to tow it.

Their route followed the Khowst to Gardez Highway, which is the major transportation artery connecting Central and Eastern Afghanistan. (See map of Rangers route above) The highway is best known as the site of numerous battles during the Afghan-Soviet War. After thirty miles of cruising on one of the best roads in Afghanistan, the platoon turned south onto one of the most treacherous tracks of terrain in the country. "The thing the map listed as a 'road' wasn't much more than an animal trail; the only sign of life we saw was some dried-up camel shit," one of the Rangers explained to me. Their route took them up and over a 6,950-foot-high ridge before dropping back down to a semi-dry creek bed that flowed through the tiny Afghan towns of Tit and Magara.

13 April, 4:00 P.M.: After an epic, all-day struggle to drive up and over the ridge, the platoon found a good piece of terrain on which they could set up a vehicle security perimeter and bed down for the night. They were within five miles of the Pakistani border and within one mile of a small Afghan military outpost known as Border Control Point #5 (BCP #5). BCP #5 was manned and operated by members of the Afghan military, whose primary purpose was to monitor and control all border crossing activity. On this day, the 2nd Platoon's company commander and company executive officer (XO) were operating out of the heavily fortified site after flying in on a helicopter a few days earlier.

For the next six days, the Rangers of 2nd Platoon used Border Control Point #5 as their operational hub. It provided them with a centrally located safe haven from which they could access most of the "grid zones" that they were tasked with clearing. The weather in this part of Afghanistan during April is temperate, with average highs of 62 degrees, and average lows of 39 degrees.

19 April: After six days of searching homes and conducting dismounted patrols throughout the area, the Rangers hadn't seen or heard about a single enemy insurgent, and had only found a couple of AK-47s, a few RPG rockets, and lots of marijuana. When we think about drugs in the context of Afghanistan, we usually think about poppy plants and heroin. As the Rangers learned on this mission, marijuana grows naturally all over Afghanistan, and along with its derivative hashish, it is used by a large percentage of the male population.

At the time (2004), many members of the non-volunteer Afghan National Army (ANA) were known to smoke marijuana and/or hashish regularly. In my own experience, most ANA soldiers we worked with seemed to recognize that using drugs before

*Marijuana grows naturally all over Afghanistan and is
used by a large percentage of the male population.*

or during combat wasn't a very good idea. Having said that, like most habitual drug users, if given access to the drug, they are likely to use it every day, and one of the most common patterns of combat is that it usually happens when you least expect it.

Afghan soldiers during "down time" at Combat Outpost Spera[41]

The Rangers of 2nd Platoon weren't the only ones who hadn't found anything of significance. None of their sister platoons from the 2nd Ranger Battalion had encountered so much as a single enemy insurgent. Nor had they gathered any actionable intelligence to assist them in finding insurgents in the future. As one of the Rangers explained it to me: "It only took one or two 'grid zones' to understand the futility of the entire concept. We were driving around one of the most rural areas of Afghanistan in slow-moving Humvees, armed to the teeth, and dressed from head-to-toe in assault gear that probably made us look like space aliens to the locals. The slow speed and high torque stress on the engines and transmission made the going extremely noisy. When you add in the clashes and clangs that occur every time a vehicle goes over a rock or crack, you realized very quickly that you have no chance of sneaking up on the enemy. Once the first few villages got cleared, the villagers spread the word to all the other villages in the area. That's what tribes do; they look out for each other."

The pointlessness of the concept was reinforced every time the platoon leader called in a grid zone as clear, even though the very fact that they were on their way to the next grid zone meant that it no longer was.

What was the logic of why the Rangers were following such an unproductive and predictable pattern of behavior? In an interview with investigators, the battalion operations officer/S-3, a major, who was one of the chief architects of the Rangers portion of Mountain Storm, explained it this way: "We went into that area and all we had was our S-2's (intelligence officer) best guess. So we drew in NAIs (named areas of interest) on the map, and the platoon went in there based off the NAIs we drew on the map. We were in there for a week or ten days and didn't have any contact. It was really a 'let's see if we could find' indicators mission.

"Throughout the operation, there were vehicle problems. Just the sheer nature of that terrain; you're going to bang up the Humvees. They're going to break. They're going to fall apart. So Humvee-part re-supply, almost nightly, was one of our main efforts. We pulled all our forces out, except that one platoon, so we could prioritize our efforts on planning our next mission, which was in a different area where the terrain is less difficult and we had better road access."

Unbeknownst to the Rangers of the 2nd Platoon, this was the logic of why their chain of command was ordering them to do what they would be doing over the next forty-eight hours. Remember it as you read.

When I first read about the Rangers' mission and their vehicle convoy configurations, all I could think of was all the rusting frames of Soviet wheeled and tracked vehicles that dotted the countryside when we first arrived in Afghanistan during the winter of 2001–2002. Initially, we assumed they were 100% the result of enemy ambushes. Over time, we learned most of them were abandoned and destroyed when and where they broke down. Reminders of one of the many hard-earned lessons the Soviets learned: that driving around enemy-occupied, mountainous terrain in armored vehicles comes at a cost; if anything goes wrong

Soviet Army vehicles on patrol (1980)

Soviet vehicle sits where it broke down, Panjshir Valley, Afghanistan.

with that vehicle, there are no local gas stations or towing services to call, and the enemy doesn't recognize administrative time-outs for maintenance in combat zones.

When a large, lumbering military vehicle breaks down in enemy-occupied terrain, there are only two options: 1) defend in place until you can fix it, tow it, or airlift it out; or 2) blow it up and abandon it. It is for this reason that during the first few months in Afghanistan (2001-2002), **we only had one rule when it came to driving around in convoys of military vehicles: Never do it.** It didn't make sense during the Soviet War, it didn't make sense in 2002, and it didn't make sense for the Rangers in 2004.

Unfortunately, the Rangers had little experience of their own that would make this lesson obvious. They also didn't have access to the foundational lessons learned by the special operations units that preceded them. As an example, in 2002, we established solid working relationships with the key warlords and tribal elders who controlled the people and the economy throughout the area the Rangers were operating in. Yet none of those key relationships or any information we learned from them was readily available and/or passed on to the Rangers. What it immediately said to me was that we (the U.S.) had, in two short years, already lost our way in Afghanistan. We had no common sense of shared purpose between units, nor any way to pass on the hard-earned lessons each unit learned while trying to accomplish it. How could we have lost our way so quickly?

> *What we do, and what we choose, is informed by the purpose, or context lens, through which we view it. To maximize the potential and performance of any type of team—combat, corporate, country, or sports—they must have a common sense of shared purpose (CSSP) and direction for accomplishing it.*

A good friend of mine, whom I served with in the Unit, and who later went to work as a security contractor for another government agency, may be one of only a handful of people who were physically present in Afghanistan for extended periods of time (2+ months) during every year from 2001 to 2021. He was one of the first in and one of the last out. Most years, he was on the ground twice a year for four to eight months.

After he returned from each deployment, we would talk about what he was seeing and learning, often while sitting around a campfire in the Bob Marshall Wilderness in Montana. When we talked in 2012, a few weeks after the Vice President of the United States guaranteed[42] we were going to pull all forces out within six months, I asked him this question: "Now that it's over, what's the biggest lesson you think the country should learn from it?" He thought about it for a moment and then explained it to me this way: **"There was no continuity of purpose. And no way to transfer the hard-earned knowledge we learned and the relationships we established along the way."**

This overarching lesson of the war in Afghanistan was echoed by every Ranger who reviewed and contributed to this book. What follows is some of their comments:

"No continuity of purpose, and no way to pass on lessons learned; that sums up the Rangers' mission in April of 2004 as well as the whole damn war." "I did three tours in Afghanistan, and no one ever told us what the U.S. purpose was for being in Afghanistan." "During the first two months of my tour in 2009, we just drove around the country and waited to get shot at."

Of note, from 2001 to 2021, our purpose in Afghanistan was driven by four different U.S. Presidents, eight different Secretaries of Defense, and at least ten different overall commanders; each one showing up with their own strategy or game plan for how they were going to turn the war around. To ensure their own success, each one brought their own cronies with them to establish and run the new leadership teams, even though most had no experience or direct knowledge of Afghanistan, its people, or its history. During the highest U.S. casualty producing years (2008–2015), the purpose/strategy in Afghanistan changed seven times (five of those in a six-month period), and that was just the overarching national/ NATO strategy.[43] All the tactics to accomplish those strategies changed with each new military unit and each new commander at each level of those units. **Each new commander and strategy also cut the cord of continuity to the lessons learned from the last. As a result, we weren't learning from our experiences, and if you can't learn, you can't adapt. History proves that if you can't adapt, you perish.**

> *When two or more brains share a common sense of purpose and the purpose makes sense, they instinctively collaborate, communicate, and reciprocate to accomplish it. Think of finding food, fighting fires, and fending off floods; think of friendships and families. Now think of twenty years of combat operations in Afghanistan.*

It only took a few years for the enemy to learn that we (the U.S.) had a learning disability. Every six to twelve months there was a new commander, a new strategy, a new military unit, and new tactics, techniques, and procedures to accomplish it. It was like Groundhog Day. As each military unit flew out of Afghanistan, everything they learned flew away with them. As a result, the Taliban realized that all they had to do to survive was wait us out.

All four of the investigations use the term "roads" when referencing the platoon's routes of travel. However, in this part of the country, most of the "roads" are actually creek beds. Creek beds are the communication and transportation arteries that connect villages

*In mountainous areas of Afghanistan, the creeks make up
the road network, and tributaries make up the side roads.*

and carry the goods and services that sustain life in and around the mountains of eastern Afghanistan. Navigating these creek beds is greatly enhanced by understanding that side roads aren't roads either. They're tributaries that empty into the creeks. As such, they are just as important when it comes to navigation. By counting and keeping track of the tributaries you've passed, you always have a good reference point for figuring out your location on the map.

The Humvee is a superb off-road vehicle. It has a low profile (six feet tall), a wide stance (seven feet), and is fifteen feet long. These proportions contribute to a stable, road-hugging truck that is very difficult to roll over. Yet those same proportions make navigating the mountainous regions of eastern Afghanistan not only more difficult but also far more dangerous. The safest passages have been blazed over the years by goat herders and drivers of small Toyota or Nissan pickup trucks that have wheelbases up to fourteen inches narrower than the Humvee's. Due to recent rains and spring run-off, there was water in all the creeks the Rangers drove through, which required extra caution and the ability of the driver to gauge the water depth directly in front of the vehicle. Often, one of the Rangers had to walk ahead of the vehicle as a ground guide to locate the safest path.

20 April: One of the platoon's Humvees broke down, and the Ranger mechanic spent the rest of the day "inside the wire" at Border Control Point #5 trying to diagnose the problem and figure out how to fix it.

21 April: The platoon spent the day at Border Control Point #5, working on their weapons and communications equipment while they continued to wait. Around midday, the mechanic diagnosed the problem as a faulty fuel pump. The platoon leader got on the radio and called Ranger battalion headquarters in Khowst to request a new fuel pump, as well as a resupply of water and MREs. Ranger battalion HQ approved his request; however, heavy rains delayed the arrival of the resupply helicopters until 11:30 P.M.

22 April, 4:15 A.M.: By the time the platoon off-loaded and distributed the supplies from the helicopters, daylight was upon them. After installing the new fuel pump, the vehicle still wouldn't start, so the mechanic told the platoon leader that he "wasn't sure what the problem is" and that he didn't "have the tools or the know-how to diagnose and fix it." *(Note: Investigators would later determine that the problem was actually a faulty solenoid.)*

When the resupply helicopter flew back to Khowst, the A company commander flew with it. Before he departed, he gave the platoon leader guidance on what he should do with the broken Humvee. The platoon leader then asked the mechanic whether he thought they "could tow the Humvee with tow straps." The mechanic told him "it was worth a try."

Once the mechanic had successfully jerry-rigged the inoperable Humvee to one that was operable using nylon tow straps, the platoon leader got on the radio and updated his company commander, who was now back at the forward operating base in Khowst. Afterward, the platoon leader told the mechanic that the decision had been made to "continue towing the inoperable Humvee since we would be going back to the FOB in Khowst in a couple of days anyway."

Taking note of the platoon's vehicle predicament, the Afghan commander at Border Control Point #5 suggested that they travel on a different route than the one they had planned the day before. He explained through an interpreter that "the route they had initially settled on would not support a towed vehicle" and that he would "provide some of his Afghan soldiers who had intimate knowledge of the roads and villages to guide them on a different route and help ensure they made it to their objective on time." The platoon leader agreed, so the Afghan commander provided a squad of seven Afghan soldiers from BCP #5 along with two Toyota Hiluxes to transport them.

6:00 A.M.: Before departing, the Rangers had to reconfigure their vehicle load plans in order to make room for the Rangers that were supposed to be riding in the inoperable Humvee (one of whom was Corporal Tillman). As a result of the reconfiguration, each of the three remaining gun-mounted Humvees (known as GMVs) would now have additional Rangers riding inside. Even with the standard load of five Rangers, it's a tight fit. When you add additional Rangers you also add additional rucksacks, weapons, communications equipment, and ammunition, along with as much water and extra fuel as possible. Once loaded, no one had any doubt about how uncomfortable the drive was going to be. Their only hope was that it would go fast.

7:00 A.M.: The Rangers drove out of Border Control Point #5 in a convoy of ten operational vehicles and one non-operational. Their destination was a town called Mana.

I asked one of the Rangers what the mood of the platoon was like when they drove out the gate of BCP #5: "After six days of driving around to random villages and knowing we weren't going to find anything, I guess we were starting to get numb to doing stupid shit. Here we were being told to tow a broken-down vehicle out of a secure base, where it couldn't give us any more problems, into enemy-occupied mountains, where it undoubtedly would."

> *One of the best ways to check and see if what you are doing "makes sense" is to say it out loud.*

Note: As mentioned above, BCP #5 was a secure base with a secure helicopter landing zone, which made it the optimal place to leave a broken-down vehicle until it could either be airlifted out or fixed on site by a more skilled mechanic. Although not mentioned in any of the investigations, the Platoon Leader requested the vehicle either stay at BCP#5 or get airlifted by a CH-47 helicopter to Khowst. Both requests were denied by the chain of command.

According to the platoon leader, "the Afghans lead the convoy on a long south-westerly route, hand-railing the Pakistan border, before eventually turning north again and moving towards their objective in Mana." They were moving through an area that had been given special emphasis in their intelligence briefings as one of the most heavily trafficked insurgent infiltration/exfiltration routes between Afghanistan and Pakistan. As one of the squad leaders mentioned before they departed, "if the platoon was ever going to randomly bump into the enemy, this route provided one our best chances, so everyone was on high alert."

The proximity to the Pakistani border, along with the fact that their convoy was being led by Afghan soldiers whom they had never worked with or vetted, made driving in broad daylight a major source of disgruntlement within the platoon. For many of the men, the source of their anxiety wasn't hypothetical; it was freshly etched in their memories from the battalion's previous tour of duty in Afghanistan.

Five months earlier (November 2003), while Corporal Tillman and his brother were still in Ranger School, the battalion deployed to Afghanistan as part of a one-month surge operation, ostensibly to assist other U.S. forces in dealing with a reported uptick in enemy activity. Once the Rangers arrived in-country they found the actual uptick to be minimal and the operational need for their specific skill sets to be none. "We were all fired up when we got there, and then we just sat on our asses with no missions, and, according to some of our buddies in other units, no real need for us to be there." It was during this time that the 2nd Ranger Battalion commander volunteered his men to escort resupply convoys of food and fuel from Bagram Airfield to Konar Province, which was 150 miles to the north.

On 14 November 2003, while escorting a resupply convoy through an area intelligence analysts had nicknamed "ambush alley," the lead Ranger escort vehicle blew a cooling hose, which forced the entire convoy to pull over while it was fixed. The men in the convoy knew from an intelligence briefing before they departed that the section of road they were driving had been the site of several recent IED (improvised explosive device) attacks. Perhaps it was that knowledge, or simply a gut feeling, that caused the commander of the convoy (a Ranger captain) to call back to Ranger battalion headquarters and recommend that the convoy "use the maintenance halt as a tactical pause to remain in their current location until the sun went down." After which, they would complete their journey under the cover of darkness and the security advantage their night vision devices provided that very few enemy possessed.

The answer they received from the battalion headquarters was immediate and definitive: "Request denied; follow the time

U.S. military Humvees driving on Pech River Road

schedule you were given in the plan." As the convoy began lining up to resume its daylight movement through "ambush alley," the mechanic was still conducting final tests on the lead Humvee, so the vehicle that had been traveling as second pulled out in front to take the lead. Not long after resuming movement, the convoy came to a one-lane stretch of road carved into the cliffs above the Pech River. The winding, narrow road forced the vehicles in the convoy to slow down and then bunch up, which cued the enemy fighters hidden in the surrounding mountains to remotely detonate a massive IED directly beneath the lead vehicle.

As described by one of the Rangers riding in the convoy[44]: *"I've never seen a Humvee so destroyed. One of our good buddies, Jay Blessing, had been driving it. He had been blown completely out of the jeep, down onto a flat area next to the river. One of his legs was all the way across the water...on the far shore. A bunch of EMTs ran down to him as fast as they could, but there was nothing they could do. It was horrible. He suffered. That was the first time I'd seen someone die. Jay was a really good guy, super-dedicated to the unit."*

Sergeant Blessing was killed by what is known as a remotely detonated IED that is triggered by a remote device such as a garage door opener or a cell phone. This enables the bomber to detonate the explosive at the exact moment the target vehicle passes over it. This method not only ensures the explosive device does maximum damage to its target, it also allows the bomber to detonate the IED from a safe location, after which he can either continue firing other weapons at the vehicles or flee from the scene without being detected or engaged.

The only way to have any chance of suppressing or destroying an enemy hiding in mountainous terrain is to have weapons that provide both the power and potency to reach out and touch them. For this reason, vehicle resupply convoys during this period in both Afghanistan and Iraq were always equipped with vehicle-mounted "crew-served" weapons such as the M2 .50 caliber machine gun, the M240 7.62 machine gun, and the Mk 19 40mm grenade launcher.

Although crew-served weapons are critical for suppressing enemy targets after they've been detected, the best way to protect a convoy from remotely detonated IEDs is to prevent the bomber from detonating in the first place. It was common knowledge, then and now, in both Afghanistan and Iraq, that almost all remotely detonated IEDs occur during daylight hours when the enemy can see the size, speed, and disposition of the target in relation to the hidden IED. **It's common sense for convoys to travel under the cover of darkness.**

As the comments that follow reveal, the circumstances surrounding the death of Sergeant Jay Blessing had a profound impact on his fellow Rangers, as well as the way the battalion chain of command operated afterwards:

"When Jay Blessing was killed, the battalion policy was night driving only. The same hurry up and get 'boots on the ground' mentality drove SOMEONE in the TOC to order the convoy commander to keep moving even though it was daylight. After Jay's death, no one seemed to remember who that SOMEONE was."

"I understood this was the kind of risk we'd signed up for. But I was really pissed off that we had been ordered to drive during the day. It was a really stupid decision, and I believe the battalion commander made the call; he was the one pushing the pins back in the TOC. It really made me question authority. I talked to my platoon sergeant about it; I talked to my first sergeant about it. Nobody would come right out and blame him because that's insubordination. In the military, you get fired for that kind of shit. But a lot of us talked about it among ourselves."[45]

"So Jay Blessing arrived in the battalion and within a couple weeks was already trying to get a slot to Ranger School. He left for Ranger School after about two months in the platoon. He went straight through even though he had a collapsed lung as a result of a parachute injury. He finished Ranger School on one lung! A few months after he returned from Ranger School, his other lung collapsed while he was doing PT [physical training]. He was rushed to the emergency room and hospitalized. All this happened within his first year in battalion. The Army gave him the option of retiring with a high disability and compensation but Jay turned it down because he wanted to stay in the Rangers. Initially, he did OK, but with his breathing/medical problems, he just couldn't keep up. Everyone admired his determination and

*work ethic, and he was well liked by all that knew him, so
he was offered a highly sought-after job working in the battalion arms room [where the battalion's weapons are stored
and maintained]. Jay grew up in Tacoma, Washington, and
would often go back to his high school—proudly wearing his
uniform and Ranger beret—to visit with students and teach-
ers and tell them how much he enjoyed being a Ranger. I
was very saddened to hear he was killed, and I have thought
about him often. He was one of America's best, and I was
blessed to know him!"*

After the death of Sergeant Blessing, there were two noticeable
changes in the way the battalion chain of command operated. The
first was a doubling down on the unwritten edict that prohibited
movement by any Ranger vehicle or convoy during daylight hours
unless they received permission from the battalion commander.
The second was described to me by a senior non-commissioned
officer who was also in Afghanistan with the battalion when Jay
Blessing was killed: "From that day forward, there was almost

Jay Blessing

no direct communications between high-ranking officers in our headquarters and platoon-level leaders in the field; everything was relayed."

This pattern of no direct communication between high-ranking leaders and subordinates emerged across the military after a number of high-profile incidents in which leaders were fired as a result of giving "improper" (aka senseless) orders to subordinates during combat operations in Afghanistan and Iraq. The lesson leaders should have learned from these incidents was: **before issuing senseless orders such as**—*continue driving down ambush alley in daylight*—**make sure you say out loud the logic of why the order does or doesn't make sense. Saying it out loud enables us to pressure test what we're thinking by using our other senses, and those of the people around us, to see whether the order makes sense or is senseless. As such, it's an organization's primary safeguard against senselessness.**

Unfortunately, some leaders took away the opposite lesson. Instead of choosing to say the logic of why out loud, they chose to say nothing at all. As you are about to learn, the implications of the "don't answer, don't tell them anything" approach to leadership can lead to tragic consequences on the battlefield.

> *The Army's emphasis on leadership is described as doctrine in FM 3-0, Operations: "The role of the leader is central to all Army operations, and trust is a key attribute in the human dimension of combat leadership. Soldiers must trust and have confidence in their leaders. **Once trust is violated, a leader becomes ineffective.**"*

With the Sergeant Blessing incident weighing heavily on their hearts and minds, the Rangers of 2nd Platoon understood that their lives depended on their ability to stay focused on the road in front of them and the hills around them. Yet with each passing moment, it was becoming excruciatingly difficult to do. Their route followed a river that snaked back and forth around blind turns, bone-jarring potholes, water-filled crevices, and beach-ball-sized boulders. Avoiding these obstacles required both intense concentration from the drivers and extreme patience from the passengers. During the first four hours of movement, the convoy drove at an average speed of 1.5 miles per hour—a pace that would have been comical if it wasn't so excruciatingly uncomfortable. **Speed is security. Lack of speed makes you a sitting duck.**

22 April, 11:30 A.M.: After four hours of struggle between the terrain and the inoperable Humvee (aka "the ball and chain"), the terrain finally delivered a knockout blow. As the convoy was passing through the small town of Magara, the tie-rods snapped and the towed Humvee collapsed to the ground. Magara consisted of a cluster of thirty-to-forty mud-walled homes that were built into the hillsides, and housed a hundred or so tribal inhabitants. When the Humvee collapsed, the Rangers stopped in the dry creek bed and did what they always do during extended halts: they dismounted and formed a security perimeter around their vehicles while the platoon leader, platoon sergeant, and mechanic discussed what to do next.

Magara, 12:24 P.M.: With collapsed wheels facing in opposite directions and the fender touching the ground, the platoon leader and platoon sergeant didn't need the mechanic to tell them the vehicle could no longer be towed. The platoon leader immediately set up his satellite radio to contact the 2nd Ranger Battalion HQ, located forty-four miles away at the forward operating base in Khowst. The platoon leader's intent was to inform his chain of command of the platoon's current situation as well as to request either a Chinook helicopter to sling-load the inoperable vehicle or a military tow truck, known as a wrecker, to load and transport the inoperable Humvee back to the base in Khowst.

Figure 1: M1089 military tactical vehicle (MTV) wrecker.

Every member of the Ranger's chain of command was located inside the tactical operation center in Khowst that day, except for the battalion commander. As he later explained it: "I was with B Company, and we could hear what was going on on the radio. So, while I wasn't physically collocated with them, I was only two or three miles down the road, maybe 15–20 minutes by vehicle."

Running the Ranger HQ staff in the battalion commander's absence was the battalion operations officer/S-3. The operations officer/S-3 is the battalion commander's principal staff officer for matters concerning operations, plans, organization, and training. He is the commander's main assistant in coordinating and planning missions. As mentioned, I was the operations officer for 2nd Ranger Battalion from 1997–1998.

The nature of the operations officer's responsibilities requires a high degree of coordination and interaction with the company commanders. Although the operations officer is a major and the company commanders are all captains, **success or failure has nothing to do with rank or the ability to order people around; instead, it depends on the relationships you establish and the trust you build with the people around you**.

The 2/75 Ranger Battalion HQ was housed in a sprawling compound of one-story, plywood-walled buildings, each with

Forward operating base in Khowst.

one or two windows and a corrugated steel roof on top. The staff officers who worked in the Ranger HQ were brought together as part of the cross-functional team concept. The CFT concept, as it was known, was the brainchild of the commanding general, who preached what he called its "twin pillars" of combat command and control. Pillar #1: Bring all commanders and their staffs together under one roof and under one commander. Pillar #2: Make every staff organization in-country look exactly the same.

The Ranger battalion CFT in Khowst received all its guidance from the Ranger regiment CFT located at Bagram Air Base. (See map.) The Bagram CFT was a massive space-age facility that was manned by over 100 staff and support personnel. The commanding general put the 75th Ranger Regiment commander and his staff in charge of running the CFT at Bagram, which meant the Ranger regiment was therefore in charge of all special operations units in Afghanistan. This was a controversial decision because the Navy SEALs were also operating in Afghanistan during this period, and these Seals are considered a "Tier One" special operations force, while the Rangers are "Tier Two."

When the SEALs protested the new command and control arrangement, their protests fell on the deaf ears of a commanding general who spent most of his career as a Ranger (he commanded both the 2nd Ranger Battalion and the 75th Ranger Regiment). In response to their protests, the commanding general told the SEALs, "Change is always difficult, but if you give this a chance, I'm confident you'll see it's the best way to operate." His guidance left no doubt that this never before used command and control arrangement was intended to be a "proof of concept" for all future special operations missions in Afghanistan.

As a result, all special operations units in Afghanistan were required to set up in the same CFT configuration. The twenty-plus staff members in each CFT sat behind tables that were arranged in a U-shape, with a large sixty-inch wide video screen hanging on the wall at the open end of the U. Ten staff officers sat along each arm of the U, with one arm designated as "current ops," and the other as "future ops." At the closed end of the U, facing the video screen, sat the operations officer as the "acting commander." On the table in front of the operations officer were his secure laptop computer and a control box that enabled him to monitor and communicate over all active radio frequencies.

The activity that dominated and drove the daily schedule of every cross-functional team across Afghanistan was the video teleconference (VTC). The purpose of the VTC is to share knowledge.

In 2004, the way CFTs shared knowledge was a top-down process. From the Pentagon in Northern Virginia to U.S. Central Command in Tampa, Florida, to the Special Operations Command in Fort Bragg, North Carolina, to Ranger Regimental HQ at Bagram Airbase, to each of the CFTs across Afghanistan, and, finally, to the guys on the ground, such as the Rangers of 2nd Platoon who had to execute it. When you say it out loud, you can see how upside-down the VTC knowledge-sharing process was.

2nd Ranger Battalion Headquarters operating under the "CFT" concept. Facility is located at the forward operating base in Khowst.

The VTCs were scheduled four to six times per day/night, with each one lasting between one and three hours, depending on the content. As a result, the cross-functional team staffs spent most of their time either preparing PowerPoint slides for the next VTC, sitting through the current VTC, or following up on whatever their higher headquarters had directed them to do in the last VTC. During the brief periods of time when there were no VTCs, the staff would eat, do physical training, take bathroom breaks, or get some sleep. During such periods, a "live" map was projected on the video screen, displaying the current location of all Ranger vehicles that were "outside the wire" conducting missions.

The live map was enabled by the presence of GPS tracking beacons known as blue force trackers. The always-on blue force tracking beacons were attached to each vehicle and provided the Ranger HQ with pinpoint knowledge of where their vehicles and people were located at any time, day or night. Like information on the Internet, it doesn't do you any good unless you log on and actively pay attention.

Behind the operations officer, underlining the U, was a row of tables where the 2nd Platoon's company commander (a captain) and his executive officer (XO) sat while they commanded and controlled the 2nd Platoon using their own radios and laptop computers. The distance between the operations officer and the company commander and company XO was six feet. The close proximity was intended to ensure that the operations officer could hear everything that was said over the radio to and from the platoons in the field.

Magara, 12:30 P.M.: After making repeated attempts to raise his company commander by both satellite radio and secure email, the platoon leader would later tell investigators, "For some reason, I was never able to talk to my company commander or the operations officer/S-3 directly. All my dealings were with the company XO, but he spoke with the authority of the company commander, so I was assuming that he was running everything through him before he gave me orders."

Note: Both the XO and the platoon leader were new to their positions. The XO's previous job was as platoon leader of 2nd Platoon. He was Pat Tillman's platoon leader in Iraq. The XO had recently been promoted to captain, so he outranked the platoon leader, who was a first lieutenant. In complex command and control situations such as this one, relationships matter and have a heavy influence on decision-making and problem solving.

When the XO answered the platoon leader's initial satellite radio request for the heavy lift helicopter or tow truck, he told the platoon leader to "stand-by" while he "checked into it." The XO then directed the platoon leader to "switch to the secure email net so we can keep the voice net clear for higher priority matters." The platoon leader complied, and for the next four hours, all conversations between the platoon leader and his chain of command were conducted via secure email. The conversations that follow are taken directly from copies of those emails.

XO: "Need a detailed description from you on what is currently wrong with the vehicle and why it cannot be towed any further."

Platoon Leader: "The front suspension is out, both the left and right sides; the front fender is almost touching the tires; it cannot be steered due to the suspension/shocks being broken; the steering box is not working as well; in addition, the vehicle will not start, and the mechanic cannot determine why it won't start (apparently it's not the fuel pump)."

Upon receipt of the email message from the platoon leader, the XO verbally relayed the message to their company commander, who was sitting in the seat next to him. How the company commander passed the information up the chain of command depended on what else was going on at the time. If a video teleconference was in progress, the company commander would either wait for the VTC to end or wait for an appropriate moment to whisper what he had heard to the operations officer.

Whenever the operations officer received an update or request from the platoons in the field, he would do one of two things. His first recourse was to contact the battalion commander and run it by him over the satellite radio to ensure he approved of what his subordinates wanted to do. If the battalion commander was not available, the operations officer would respond with his own guidance. Since all of the conversations between the battalion commander and the operations officer were conducted on a secure satellite radio, the only record of what was said between them was what was passed back down to the company commander, who then passed what he was told to the company XO, who would then type it up as an email and send it back to the platoon leader on the ground in Magara.

This was the "revolutionary way" all the CFTs across Afghanistan were organized to command, control, and make decisions for their teams on the ground in April of 2004. By saying it out loud and reflecting on the decisions the CFTs made through the reality-revealing lens of hindsight, we are better able to understand both how convoluted the process was, as well as its inherent inability to make sense of the complexities of combat, or make sensible choices about what the men on the ground should do next.

The lesson and historical takeaway from the CFT concept is one the military had previously learned too many times to count: that **hierarchical decision-making by disconnected[46] chains of command never has, and never will make sense in situations that entail any degree of complexity**. The higher the degree of complexity, the more senseless a detached hierarchy gets. The logic of why is self-evident. The purpose of our nervous system is to make sense of what's going on around us in order to make sensible choices about what to do next. The only way our nervous system is able to make sense of anything is by paying attention with our senses (sight, sound, smell, taste, and touch). Detached hierarchies, by their very definition, are "not attached" to the environment of which they are trying to make sense. When our senses are detached from what's going on around us, we are senseless.

Technology deludes the detached hierarchy into a false sense of omniscience. Eyes in the sky (satellites and surveillance aircraft) provide a narrow, one-dimensional view of reality. Satellite radio technology provides sound bites that may or may not reflect what the person speaking actually means, and laptop computers crunching numbers on technical variables, such as fuel levels and payloads, produce comments like, "We need to hurry up on the target or the helicopters are going to be running low on gas."

The higher an individual is in the hierarchy, the more detached they are from the reality of the situation. If a leader's brain can't make sense of what is happening, they have zero chance of making a sensible choice about what to do next. The only way a detached hierarchy can have any chance of making sense is by constantly deferring to the reality-revealing context provided by the guys on the ground *who are actually touching the environment.* By nature, the larger and more rigidly entrenched the detached hierarchy, the less capable they are of paying attention to the guys on the ground, and the more senseless the choices they make about what to do next.

The second reason detached hierarchies don't make sense is because they aren't scalable. Scalable means "capable of being easily expanded or upgraded to take on bigger or more numerous demands."

"Bringing all commanders and their staffs together under one roof and under one commander" may have sounded good to the commander who came up with the CFT concept because it makes it easier for him to receive reports and pass guidance. However, by putting all your sub-element commanders and their staffs under a single commander and his staff, you limit the ability of the chain of command to scale up and deal with more than one issue, problem, or opportunity at a time.

The brain can only think of one thing at a time, and the commander at the top of the chain can only focus on one situation and make one decision at a time. As an example, if two or more platoons need support or approval at the same time, their requests have to get in line as the hierarchy passes the information up the chain. In this case, from platoon leader to XO, XO to company commander, company commander to operations officer, operations officer to battalion commander, battalion commander to Ranger regimental commander at Bagram, and, finally, Ranger regimental commander to the special operations commander at Fort Bragg.

As the information passes up each level of the hierarchy, the platoons' requests are not only competing against each other for attention, they are also competing against all other activities that the CFTs prioritize, such as participating in an on-going VTC with higher headquarters or preparing for the next VTC. Not only does the detached hierarchy take longer to make decisions, the decisions it makes rarely make sense.

When it comes to complex decision-making and problem solving, detached hierarchies aren't attached to the environment and are therefore unable to make sense of what is happening on the ground. They are also not scalable, which renders them incapable of dealing with more than one complex decision at a time. The larger and more deeply entrenched the hierarchy, the less capable they are of making sense.

> *There is something exceedingly ridiculous in the composition of monarchy; it first excludes a man from the means of information, yet empowers him to act in cases where the highest judgment is required. The state of a king shuts him from the world, yet the business of a king requires him to know it thoroughly. Thomas Paine*

Magara, 1:30 P.M.: One hour after the platoon leader requested either a Chinook helicopter or wrecker to pick up the inoperable Humvee, he finally received an email response from his chain of command in Khowst:

XO: "Bottom line is that you need to get that vehicle further north so we can get a wrecker to pick it up, we will take the wrecker as far as we can take it, but everyone here is adamant we cannot take it very far off the Khowst-Gardez highway. If we try to sling it [with a helicopter], it will take 2–3 days to get air to you. I think the best option is to get you the parts you need to tow it to the highway."

Platoon Leader: "Roger, as you can see, there is no easy answer, the only other option I can think of is jerry-rigging a tow somehow with a winch and dragging it. We're wargaming it right now. We are willing to try it for sure, but there's a lot of earth between us and the highway [12–15 kilometers]."

While the Rangers brainstormed around the broken vehicle, the people from the village grew increasingly curious about the unannounced visitors, who didn't seem to be in a hurry to go away. Most of the villagers were now standing outside their homes, watching cautiously while keeping their distance and sizing up the Americans. Children were the first ones to make contact, smiling and sticking out their hands in hopes of being the benefactors of the generous bounty that the receptive Rangers immediately bequeathed. MRE crackers and candy are favorited food items of most kids, and these were no different, devouring the gifts almost as soon as they received them. As so often happens, when one kid gets something good, it emboldens all the others to try for themselves. Within minutes, almost every kid in town was swarming around the vehicles, which reassured the adults, who slowly but surely followed their lead.

Afghan children approach U.S. soldiers in Khowst Province, Afghanistan

Like the people in most other villages that the Rangers had encountered, the people of Magara seemed genuinely friendly. As one of the Rangers described it to me: "Kids are the best indicators of what the adults are thinking. No matter where in the world you are, as the kids go, so too do the adults. I always pay real close attention to the kids. If they act skittish or afraid I get my guard up. These kids were happy to see us, and for most of the day I was pretty sure most of the adults were OK with us too."

As the day wore on, many of the Rangers took note of what they described to investigators as "a handful of fighting-age men" that kept their distance from the rest of the villagers while they sat and stared at the Rangers from a house on a nearby hilltop. "You can pick up a vibe from people if you're around them long enough, and after the first few hours, these dudes' facial expressions and fidgety behaviors left no doubt in any of us that they weren't friendly. They were giving us the stink eye, so I never took my eye off them for more than a few seconds at a time."

The men on the hill provided a reinforcing reminder to the Rangers that **although good relations with the locals is always paramount, you can never let your guard down and must maintain situational awareness of your surroundings.** Based on their location (parked on the edge of the village), as well as the hundred or so residents who were now milling about in their midst, it was becoming increasingly difficult for the Rangers to do.

It was during this time that the Rangers' interpreter began talking to a local Jinga[47] truck driver, who offered to assist the Rangers by towing their Humvee to wherever they needed to go for $120 U.S. dollars.

The platoon leader agreed to the offer, and the Ranger mechanic immediately went to work jacking the front end of the Humvee off the ground so they could chain it to the back of the Jinga truck. The platoon leader then sat back down in his vehicle with his mini-laptop computer to continue his discussion with the XO and "try to figure out what to do about our Mana mission."

2:30 P.M.:

Platoon Leader: "It looks like the jerry-rigging thing may not work…a local Jinga truck driver said he could tow our vehicle with his truck and take it with us to the highway for $120. If the Jinga doesn't work, we'll have to come up with something else."

XO: "OK, I need to know asap if this system you are rigging is going to work. **I don't care what happens to the vehicle, just try to get it as far as you can.**"

Word spread quickly amongst the men of the platoon; that battalion headquarters had turned down their request for a heavy lift helicopter to extract the vehicle. Airlifting the Humvee seemed like a no-brainer to the Rangers of 2nd Platoon. A few hours earlier, during their drive to Magara, they had watched as four CH-47 Chinook helicopters flew over them on their way to Khowst. Later, the investigators were unable to locate any record of a request made by battalion headquarters for additional aviation support. Investigators did discover that the battalion headquarters in Khowst already had a Chinook helicopter dedicated to them for use that day but didn't consider providing it to the platoon because it was scheduled to take the battalion commander somewhere else later that day.

With their request for a Chinook helicopter denied, almost every Ranger interviewed (including Kevin Tillman) told the investigators that the only option that made sense at this point was to "pull the .50-caliber machine gun from the Humvee's turret, disconnect its radios, and then blow the damn thing up with enough C-4 explosive charges to make it completely unsalvageable." What was the logic of why the Ranger chain of command never even considered it?

Investigator: "Why did the chain of command order the platoon to continue dragging the Humvee instead of blowing it up and leaving it?"

Battalion Commander: "That goes against Army policy…we don't ever do that because these vehicles then would end up in the hands of the enemy people. They get on them. They take pictures. And, in that country, everything is about power and communications. And the stories that would happen after that ambush, they would take the pictures. They would say—or, they would take pictures on it. They would put the enemy on it, and this kind of stuff is probably used for propaganda purposes…the bottom line is that, in the Army, it sometimes goes all the way up to the Secretary of the Army. And I gave them the example of a tank that was destroyed in OIF [Iraq 2003] and some of the trouble we'd been through to try to recover it."[48]

What follows is some additional clarifying information I discovered while trying to make sense of the chain of command's decision to tow the Humvee instead of blowing it up.

There was no Army policy in 2004 against destroying vehicles in combat, nor is there any such policy today. As far as "the enemy using the vehicle for propaganda after the ambush," as you'll learn in the pages that follow, there never would have been an ambush or a friendly firefight if the platoon hadn't been forced to split up so they could continue towing the inoperable Humvee.

As for the "tank that was destroyed in OIF" (almost exactly one year to the day earlier), the tank and its crew had been attached to a special operations task force known as Task Force Wolverines.[49] The Wolverines consisted of members of the Unit as well as one platoon of five M1A2 tanks, and a large contingent of 1st Battalion Rangers. The recommendation to destroy the tank was made over the radio by Lieutenant Colonel Mike Kershaw[50], a highly respected leader and highly decorated former tank commander who at that time was serving as the 1st Ranger Battalion Commander. I know what he said because I was the commander of the Wolverines and he was making the recommendation to me.

What was the "logic of why" he recommended we blow up the tank? 1) The tank had flipped upside down into a ditch twenty feet deep (see picture). 2) After inspecting the terrain around the ditch, the tank crew assessed that recovering it would require a semi-truck wrecker and a crane, along with their crews, which would take two to three days of travel to arrive at the site and an additional two to three days to lift the tank out. 3) The ditch that the tank was wedged into was located just outside the enemy-infested city of Baji, Iraq, where, only a few days earlier, other members of the Wolverines had been attacked by Fedayeen[51] fighters. The risk involved in leaving a sufficient number of Rangers to guard the tank for two to six days while waiting for the recovery vehicles and their crews far outweighed the potential benefit of salvaging it.

M1A2 tank flipped into a ditch during a night attack outside Baji, Iraq. The crew escaped via the hatch on the right, after which the crew and ground force commander recommended we destroy the tank in place, which I approved. The hole on the left is from a Sabot round.

I approved the recommendation to destroy the tank without seeing it and without hesitation because it made sense for the mission, it made sense for the men, and Mike and the tank crew's explanation made sense to me. To render the tank inoperable, the tank crew disconnected all sensitive equipment (weapons, radios, night vision/target acquisition devices, etc.) and then used one of our four remaining tanks to fire a Sabot round[52] through the belly of the otherwise impenetrable beast. Although the monetary value of the M-1A2 tank was, at that time, somewhere north of $4.5 million, the cost of the tank never came up in our discussion. No one in the Army or special operations chains of command ever challenged our decision or admonished either of us for destroying the tank and protecting our people.

For past, present, and future leaders, the common sense lesson is the same: **You'll never lose sleep or second guess yourself for choosing to destroy a vehicle for the safety and survival of your people; conversely, you may never forgive yourself and will always second guess yourself if you choose to save a vehicle and end up getting any of your people wounded or killed because of it. When the situation going on around you warrants it and/or the guys on the ground share the logic of why it makes sense, use common sense and destroy the vehicle.**

Note: During the disastrous withdrawal from Afghanistan in 2021, the U.S. government made the decision to leave $85 billion worth of weapons and equipment in Afghanistan, including somewhere between 7,500 and 10,000 fully operational Humvees.

Magara, 3:00 P.M.: With the platoon sergeant and squad leaders hovering nearby, all eyes were now on the platoon leader as he hunched over his Labretto mini-laptop computer and pounded out questions and answers to the XO concerning what the chain of command wanted the platoon to do next.

According to the email records, they discussed three options:

1. Split the platoon into sections—send one section with the disabled vehicle and the other to the objective in Mana.
2. Send the whole platoon to the highway, drop off the disabled vehicle, and divert back to Mana.
3. Cancel the mission and send the whole platoon with the disabled vehicle back to the FOB in Khowst.

The XO told the platoon leader that he would get back to him as soon as he had an answer. While the platoon leader waited patiently, the rest of the platoon was beginning to lose theirs. With each passing minute, as the sun dipped lower in the sky, the security situation around the Rangers was growing more and more precarious.

According to the platoon sergeant, "Once the platoon was in town trying to rig the vehicle, all Rangers were involved pulling security due to the large amount of people. The main concern while rigging the damaged vehicle was that no locals tried to take something from the other vehicles." This was not an ancillary detail. In addition to guns, ammunition, and assault gear, their secure radios contained top-secret security hardware, which the military considers so sensitive that the use of deadly force is authorized to protect it.

I asked the Ranger private who worked for Corporal Tillman what they did while in Magara.

Ranger Private: "All our time in Magara was spent waiting on the word. Most of us sat, or kneeled, or stood in the same spot for **almost five hours** while pulling security. Even though we were supposed to be watching the locals, most of us were focused on listening to the platoon leader's conversation over the radio. Initially, everyone thought we were blowing up the vehicle, then we heard they denied our request to blow it and told us to tow it instead.

'That can't be right,' I thought, but then again, I'm only a private. Then we heard we were supposed to tow it with a Jinga truck and local driver—now we were worried—we were hanging on every word. It was a roller coaster with mostly lows and no highs; 'This isn't good,' was what I was thinking.

"As for Pat, when he wasn't huddling with the squad leaders and team leaders, he was checking on me and the guy next to me: 'Do you guys have enough water, do you need something to eat?' He always told me that we needed to stay focused on our surroundings not on our stomach so he was constantly reminding me to eat and to stay alert, and stay focused."

The newly appointed regimental sergeant major, who was riding along with the platoon, was also growing concerned. After taking the job as regimental command sergeant major a few weeks earlier, his first order of business was to spend time with every platoon in the regiment. The 2nd Platoon was the first of thirty-six platoons he was scheduled to ride along with. The choice wasn't completely random, he started his military career as a private in A company, 2nd Ranger battalion from 1978 to 1986. He then spent fifteen years as a member of the Army's premier special mission unit where he deployed all over the globe on some of our nation's most sensitive missions. He returned to the Ranger Regiment in 2002 and had an immediate impact on the organization and the mission when he earned a Silver Star for his leadership under fire at Haditha Dam during the first few weeks of combat in Iraq. As one of the most experienced and highly respected combat veterans in the Army at the time, the men of 2nd Platoon were extremely appreciative of his presence.

More than one Ranger mentioned the calming influence the regimental sergeant major had on them: "Our morale was low when we left BCP #5, so by the time we got to Magara, it was lower than whale shit. **Having the regimental sergeant major with us**

made us feel like somebody outside our own platoon actually cared about us." Another Ranger recorded in his journal, "He did everything we did, he pulled his shift on guard duty, helped clean the crew-served weapons, and we all knew he didn't have to do any of that, he didn't even have to be out there with us."

After asking the platoon leader how he was doing, the regimental sergeant major shared his concerns about the security situation: "We shouldn't be sitting around this long, Sir," he advised, and then added, "We need to get moving." The platoon leader wholeheartedly agreed. Every Ranger on the ground shared a common sense of urgency to get out of Magara. Unfortunately, it was a sense that wasn't shared by their chain of command, who were making all the decisions from the comfortable confines of their forward operating base forty-four miles away in Khowst.

This is why the "context of the moment" is such an important ingredient for making decisions and solving problems. Without experiencing the context of the moment—think discomfort, security concerns, unfamiliar people milling around the vehicles, the handful of fighting-age men on the hill, and the sun dipping below the horizon—our brains can't make sense of what's going on around us or make sensible choices about what to do next.

1) What's going on around us in the context of the moment + 2) What we know = 3) How our brains make sense and sensible choices about what to do next.

1 + 2 = 3, To make sense, it has to add up.

Magara, 4:00 P.M.: One hour after the platoon leader requested guidance on what they were supposed to do next, and three and a half hours after their initial request for assistance, the platoon leader received an email answer back from his chain of command.

XO: "Here's what needs to happen…

1. Let me know when you SP [depart] with the Jinga and broken vehicle.

2. Send a section along with both mortar vehicles back to the K-G highway. BOTH mortar vehicles, the 81 tubes, and mortar personnel along with the SNIPERS need to come back to the FOB...so send them with the broken vehicle. Keep your 60 mm mortar section with you.

3. Send one section into your next objective area to start clearing. The escort section will then meet you in zone after they drop off the broken vehicle.

4. A wrecker with a GMV [gun-mounted vehicle] section from 1/C [1st Platoon, Charlie Company] will link up with you on the Khowst-Gardez Highway [grid redacted] they are linking up at 17:30 and will SP shortly thereafter...How copy?"

According to one of the Rangers who was sitting near the platoon leader but had no idea what the email said, "We could tell he got an answer that nobody wanted to hear. When he looked up and realized we were all watching him, he said, 'I'm trying, guys, I'm trying as hard as I can.' Then, almost as if he caught himself, he put his head back down and started typing another message. He was a really good officer and we could tell he was having a hard time because it's really rare for an officer like him to share what he was feeling with a couple of enlisted guys."

The platoon leader was having a hard time making sense of what he was supposed to do because this was an order with no logic of why attached to it. The logic of why provides our brains with the logic of purpose. It informs why we do what we do and why we choose what we choose. **Why is it mission-essential for all leaders to communicate the logic of why?**

When striving to understand the choices we make, it's not the choice itself (yes, no, buy, sell, etc.), it's the "logic of why" that we used to make the choice, which explains, communicates, and validates whether the choice makes sense or is senseless. The logic of why provides leaders with three built-in checks and balances that help expose misguided commands, orders, and choices before they see the light of senselessness:

1. Say the logic of why out loud and/or write it down to make it physical so our other senses and those of the people around us can pressure test it and see if it makes sense. The more senses we involve, the more sense we can make.

2. Ensure the logic of why consists of three or more context-specific patterns and/or time-tested principles that we can see, hear, smell, taste, touch, feel, and make sense of. First of all, secondly, and thirdly…If it doesn't have three legs to stand on, it's probably not standing on solid ground.

3. Ensure the logic of why coheres with our common purpose (e.g., to clear Mana) and adapts to what's going on around us in the context of the moment (time of day, security situation, status of men and equipment, etc.). A coherent-adaptive choice is a sensible choice (keep the platoon together and clear Mana the next day).

In situations where time is of the essence, the logic of why can be as simple as "I've got a gut feeling about this." Context is everything, so if the leader telling you they have a gut feeling is trusted and the situation warrants it, you may say "good enough" and go along. The key is that you now understand that the logic of why you are doing what you are doing is based solely on your leader's gut feeling. If an incoherent pattern pops up that dispels the gut feeling, you can self-correct and try another way or go back to the drawing board and come up with something new. Freedom of choice to change our minds is common sense.

Magara, 4:10 P.M.:

Platoon Leader: "I would recommend sending our whole platoon up to the highway and then having us go together to the villages, and here's my reasons:

1. Splitting into sections would mean that one of the sections only has one gun truck with it, since the other one is inoperable.

2. With the waning daylight, by the time the objective section reaches the objective in Mana it's going to be dark. Clearing the villages at night, given our current TTPs[53] [tactics, techniques, and procedures], seems like it would not work.

3. I'm the only one with a satellite radio. Mortars are returning to Khowst, if something were to happen to the highway section, they'd be in a comms void.

"I recommend that the whole platoon escorts to the highway, sets up for the night just north of the first village, and begins clearing south when the sun comes up. If this does not fall in line with your intent, we will execute as previously discussed."

The platoon leader's response provides a textbook example of the logic of why it made sense to keep the platoon together and clear Mana the next day. The purpose that the Ranger chain of command gave them was to have "boots on the ground outside Mana by nightfall," a purpose they could have easily accomplished if they had been allowed to move there together as one platoon. Once outside Mana, they could have set up a security perimeter and gotten some sleep (the battalion had a standing policy against searching villages at night), before waking up at sunrise, meeting with the village elders, and then completing their mission of "clearing" Mana (which would turn out to be yet another dry-hole). After clearing Mana, they could have departed the area together and continued following Canyon Road directly north for five kilometers (see map) to link up with the wrecker near the K-G highway. All of which could have been accomplished before noon the next day.

Magara, 4:35 P.M.: At this point, the platoon had been in Magara for five hours. Four of those five were spent waiting on email answers from their chain of command.

> *Information is withheld until the last possible momen,t*
> *with little concern as to how it will affect subordinates.*
> *"Heads up" or "Warning orders" to initiate activities*
> *and movements aren't sent out until the last minute.*

After sending his objections to the XO regarding the order to "split the platoon," the platoon leader called in his platoon sergeant to let him know what they were being ordered to do.

Platoon Sergeant: "I didn't want to do it…and I said, 'Sir, this doesn't make any sense, you know, whose idea is this?' and he was like, 'it's from higher,' so I left and came back and my RTO [radio operator] was standing there so I tried to stay calm because I was pretty aggravated, and I said, 'this makes no sense.' I said 'who's telling you this.' And he said, 'I've got an email,' and I said, 'I'm not calling you a liar, but I want to see the emails,' and he said, 'hey, I'll call again,' and he got on the radio. He seemed like he was pretty aggravated about it…that's why when I left and came back, the way he looked, he looked like he was aggravated, like he had tried all he could and it still was—we were still doing it."

Investigator: "Was there any way to go higher than your company chain of command to protest the decision to split the platoon?"

Platoon Sergeant: "No. When the order came down from higher headquarters there is no protesting the decision. He voiced his concerns to the XO who was the only person he could talk to, and the end state was made by HQ to split the platoon."

No direct communication with subordinates. Everything goes through the hierarchy/staff or electronic messaging.

The platoon sergeant would later tell investigators that it was the maddest he had ever been with his platoon leader and that they had a pretty good relationship, but the platoon leader was splitting the platoon based on higher guidance. And he (the platoon sergeant) was dead set against that.

Based on the statements made by those around the platoon leader that day, it was obvious that the stress of going back and forth with his unresponsive chain of command had begun to take a psychological toll on him. The regimental sergeant major summed up what he observed to the investigators, "He was professional, and he was overwhelmed."

The inability to speak up, out, or with leaders to ask clarifying questions leads to confusion amongst subordinates, continuous uncertainty, and stress.

The second investigation (conducted by Ranger regimental HQ) made note of "the platoon leader's lack of experience as a contributing factor" to the sequence of events and the firefight that followed. First lieutenants typically have between eighteen and thirty-six months of active duty experience, so, by proxy, almost all lieutenants technically "lack experience." Yet, the platoon leader's objections to the "split the platoon" order weren't based on his lack of experience. Instead, his objections were based on what was going on around him and his men in the "context of the moment" in Magara. As the Rangers around him astutely observed, "he was stressed out, and he was crying out for help."

> *The conflicts that cause the most stress in combat are the contradictions between what our order, plan, or disconnected chain of command tells us to do, versus what the reality of the situation reveals we should do.*

The platoon leader was stressed because his brain couldn't make sense of the order his chain of command was telling him to follow. The only way to reconcile these types of conflicts and the stress they incur is to communicate the logic of why the order does or doesn't make sense. Yet none of the leaders in the Ranger chain of command were available to talk with him on the radio and listen. Therefore, the conflict between what he was told to do by his disconnected chain of command versus what the reality of the situation going on around him revealed he should do continued to grow—as did the stress it produced.

Magara, 5:05 P.M.:

Frustrated with the lack of answers he was getting from the XO, the platoon leader directed his radio operator to switch back to the battalion's satellite radio frequency in the hope it would enable him to talk directly to his company commander or the operations officer/S3 so he could share the logic of why the order to split the platoon didn't make tactical sense.

Once the satellite radio was on the proper frequency, the platoon leader attempted to contact his company commander: "Alpha-6, this is Alpha 2-6 over." No response. He tried a second time, and the XO answered: "This is Alpha 0-5 over." The platoon leader was disappointed but not dissuaded because he knew that all other key members of his chain of command (such as the battalion operations officer, the battalion commander, and the Ranger regimental commander, along with all of their key staff officers) monitored the battalion's satellite radio frequency around the clock. Consequently, they would overhear his objections to the "split the platoon order" and hopefully intervene or overrule it. Unlike the email conversations above, there is no physical record of this conversation. What follows comes from the sworn statements provided by the platoon leader and XO during interviews with investigators:

Platoon Leader: "Did you receive my message about why I didn't want to split my platoon?"

XO: "Yes, you still need to execute the first course of action."

Platoon Leader: "Is your intent for me to start clearing the village at night?"

XO: "No."

Platoon Leader: "It doesn't make sense to split my platoon just so one section could set up a vehicle drop-off point for the other."

XO: "You still need to move as planned."

Platoon Leader: "So the only reason you want me to split up is so I can get boots on the ground in sector before it gets dark?"

XO: "Yes."

Platoon Leader: "Even though we only have one fifty-caliber machine gun between both sections, does that change anything?"

XO: "It doesn't."

> *Orders, commands, and directives are issued without the logic of why they make sense. "Because I said so," "Just do what you're told," and "You don't have a need to know" are common refrains when a subordinate asks why.*

The platoon leader waited a few more minutes in the hope that someone else in his chain of command would speak up, out, or with him. Instead, all he heard was the silence of consent. The entire chain of command above him chose to say nothing.

Platoon Leader: "At that point, I figured I had pushed the envelope far enough and accepted the mission. **With the sun dropped below the ridgeline, I assessed that my section needed to move with haste to accomplish the intent of getting to Mana before dark.** I gathered the squad leaders, told them how we were going to be task organized and why, and went over the plan briefly for the rest of the night and the following day."

According to the third investigation: "Understanding but not agreeing with his orders, the platoon leader commenced hasty troop leading procedures (TLPs) in preparation for the movement." Based on the specific orders he was given regarding who was supposed to return to Khowst (e.g., "BOTH mortar vehicles, the 81 tubes, and mortar personnel along with the SNIPERS need to come back to the FOB"), the platoon leader task organized the remaining members of the platoon into two sections he called "serials."

Serial #1 was comprised of sixteen Rangers riding in four vehicles (one gun-mounted Humvee, one cargo Humvee, and two Toyota Hilux King Cabs), and seven Afghan Military Forces riding in two Toyota Hiluxes (six vehicles in total).

Serial #1 (16 Rangers, 7 Afghan Mil)

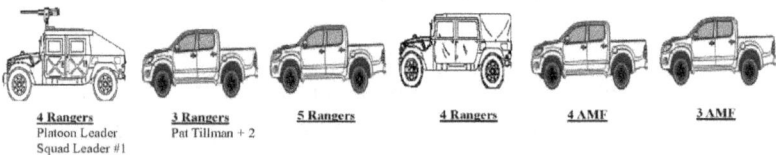

4 Rangers	3 Rangers	5 Rangers	4 Rangers	4 AMF	3 AMF
Platoon Leader	Pat Tillman + 2				
Squad Leader #1					

Serial #2 (19 Rangers, 2 Afghan Civ)

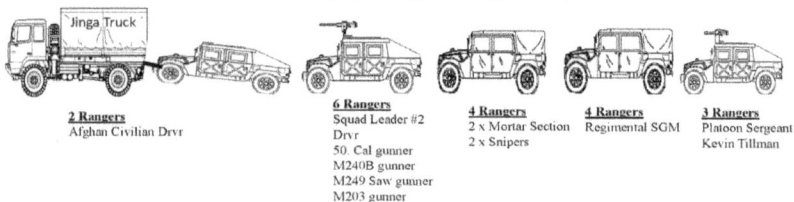

2 Rangers	6 Rangers	4 Rangers	4 Rangers	3 Rangers
Afghan Civilian Drvr	Squad Leader #2	2 x Mortar Section	Regimental SGM	Platoon Sergeant
	Drvr	2 x Snipers		Kevin Tillman
	50. Cal gunner			
	M240B gunner			
	M249 Saw gunner			
	M203 gunner			

Jinga Truck

Serial #2 comprised nineteen Rangers riding in four vehicles (two gun-mounted Humvees and two cargo Humvees) and two Afghan civilians riding in an Afghan Jinga truck towing the inoperable Humvee (six vehicles in total, one inoperable).

The two serials would conduct movement on separate routes. (See map, page 168) Serial #1 would follow a westerly route (Margarah to Mana), while Serial #2 would follow a long, winding northern route (from Margarah to Tit, then Afzal Kheyl, and finally Ragay).

After ensuring the squad leaders understood the task organization for both serials, the platoon leader and platoon sergeant went over the internal communications plan for the two serials.

Platoon Leader: "Given the distance that would soon be between us, we would not be able to talk on our platoon net for long, and should the towed Humvee become further disabled or they take contact, he would have to talk on the satellite radio. I also told him that when they got close to our platoon link-up location near the OBJ [Mana], to contact us on our platoon net. Finally, I told him that since his section was going to be slow-moving with the towed vehicle, and I had a deadline to meet, that I was going to move out with my section quickly and for him to move out with his section when he was ready."

After the platoon leader answered questions about the task organization and communications plan, the squad leaders went back to their squads to pass on the new information. According to the second squad leader (hereafter referred to as Squad Leader #2), it didn't go over well with his team.

Squad Leader #2: "OK, listen up guys, the platoon is hiring a Jinga truck from the village to tow the busted-up vehicle. The platoon will be split into two elements. Serial One and Serial Two. Serial One will be commanded by the PL and is going to clear an objective called Mana tonight. We're assigned to Serial Two. Our job is to escort the Jinga truck and Hummer back to the hardball road."

Ranger Machine-Gunner: "You're kidding right, sergeant, we're splitting the platoon just so we can get a busted-up vehicle to the hardball? What's so special about Mana?"

Squad Leader #2: "Nothing in particular. There's no intel on it. We were supposed to clear it days ago, but the broken vehicle delayed us, and the chain of command is upset. They want it cleared asap."

Ranger Machine-Gunner: "Our combat effectiveness will be cut in half so some of us can get to a meaningless objective on time and the rest of us can get that piece of shit to the highway?"

Squad Leader #2: "I'm not sure why we're doing it, especially the part about driving there before it gets dark. I recommended we move out of Magara and then wait another hour to drive to our objectives when it's pitch black."

Radio operators tend to know more about what is happening than other members of the platoon, as they are privy to all incoming and outgoing communication traffic, as well as most of the face-to-face conversations that occur as a result of those communications. They also understand the context in which communications are received and delivered.

Platoon Radio Operator: "Nobody on the ground in Magara thought it was a good idea to split the platoon; the platoon leader didn't want to do it. The platoon sergeant was totally against it, as were all the squad leaders. But in the Army, you obey orders. If somebody with a higher rank tells you to do something, you do it. So the platoon leader split the platoon."

Purpose and priorities are often ambiguous, and
when they are present, they rarely make sense.

Regimental Command Sergeant Major: "I think the platoon leader went through all procedures properly. He called higher, and when he got a decision that he didn't agree with, he respectfully questioned it and tried to explain the circumstances where he was at. And when he was told otherwise, he just sucked it up and completed what he was instructed to do with Serial One and Serial Two moving separately."

When asked by investigators about the order to split the platoon, the regimental command sergeant major responded matter-of-factly, "I think it was a bad decision."

Even though no one on the ground agreed with the order to split, most agreed it was in their best interest to leave Magara as quickly as possible. With less than an hour of daylight remaining, the activity in and around the vehicles became frenetic.

During the third investigation, the Alpha Company first sergeant, who was monitoring the operation from the forward operating base in Khowst, was asked by investigators: "What necessitated in this mission right here that they had to get down there so quickly?" The Alpha Company first sergeant provided a timeless answer: "I don't think there was anything, I think that a lot of times at higher headquarters—maybe even, you know, higher than battalion headquarters—they may make a timeline, and then we just feel like we have to stick to that timeline. There's no—there's no 'intel' driving it. There's no—you know, there's no events driving it. It's just a timeline, and we feel like we have to stick with it; and that's what drives that kind of stuff."

Timelines themselves don't drive the behavior of those who have to follow them. The behaviors are driven by the choices their leaders make and the actions they take to enforce them.

What was the logic of why the chain of command ordered the platoon to split? As stated earlier, determining "who said what" is not the purpose of these pages; the purpose is to learn from it. What future leaders should focus on is not who but why. The platoon leader already explained the logic of why splitting the platoon didn't make sense. As you read through the chain of command's sworn statements to investigators, ask yourself if any of them explain the logic of why the decision to split the platoon made sense.

Investigator: "Who made the decision that the platoon should split?"

Operations Officer: "The platoon leader, Sir. And this is where—this is consistently—I've been interviewed on this a couple of times. We developed a plan for—this is the plan as we developed it. And, actually, the irony of this whole thing is I felt very comfortable with the plan because with me when we planned it was the company commander and the XO. I mean, they were back there with me. The battalion commander called them back to begin planning for the next operation."

Investigator: "Why was the platoon movement split into two convoy serials? Did you approve of the plan?

Operations Officer: "I did not order the splitting of the platoon. Our battalion commander was pushing us pretty hard to 'keep moving,' but that was not surprising. My early comment to the company commander was, 'this vehicle problem better not delay us anymore.' We were already delayed twenty-four hours. The decision was foolhardy. Divided in two, they didn't have enough combat power to do that mission."

Investigator: "Who made the decision that the platoon should split?"

Company Commander: "We were in the operations center. We were kind of working through the decision process on getting this vehicle. We knew we had to get it back. We couldn't leave it in the zone. We also did not think there was any enemy in this area, so moving it out was just confirm and denial. We expected to deny and then move back to start the next operation. I was having a discussion with the operations officer, who was the CFT commander. He told me, 'We cannot'—I don't know exactly how he worded it. Words to the effect, 'Hey, we can't have one vehicle stopping an entire platoon,' and that we needed to send one section of the platoon forward into zone and move one back with the broken vehicle to link up with the wrecker. My understanding after that was that he said to split the platoon. He may or may have not said that. That might have been the way I understood it. So I gave the order through my XO to have the platoon move one section into their assigned zone and one section with the disabled vehicle."

Investigator: "You don't recall what he said to you last?"

Company Commander: "I know he said that. I think he said we would need to look at splitting the platoon or something to that effect. He did not direct me to split the platoon; that was ultimately my decision. The other reason I wanted to split them was like I said, Sir, we were trying to get the platoon back for planning. We were done with this area. We were trying to get them back and save them for the next part of the fight. I knew that we were messing around with this vehicle, and it was a possibility that we're falling behind."

Note: During the first investigation, the company commander, who was a captain, told the lead investigator (also a captain) that the operations officer told him to split the platoon. However, during the second investigation, the company commander told the lead investigator (a lieutenant colonel from Ranger Regimental Headquarters) a different version of who came up with the order.

A Company First Sergeant: "I had walked over to the TOC there in FOB Salerno just to find out the status of our platoon. 'Hey, where are they at and when are they coming back in to refit?' because we were all kind of rotating through, refitting the different platoons. And, as I was in there checking on it, they were giving them the movement orders, and realized they had the vehicle down and all of these things; sorting out what these guys were facing on the ground there. I remember the company commander walking over to me saying, 'They don't want to split the platoon.

They want to keep them together.' And I said, 'I don't think they should. I think they should keep them together.' And he says, 'I'm going to go talk to the operations officer/S-3 again.' [The Company Commander] didn't think they should be split. I didn't either. So I assumed he went and expressed that he didn't want the platoon to have to be split. But they were told by the S-3, 'We're not going to let the truck keep us from getting into our next village.'"

I talked to one of the CFT staff officers (a captain at the time) who was present during the conversation between the operations officer and the company commander, and this is what he told me:

Ranger Staff Officer: "I was sitting in my seat at the U-shaped table when the operations officer stood up from his chair, slapped his hands on the desk, and told the company commander, 'Tell your platoon to split.'"

"Why do you think he wanted them to split?"

Ranger Staff Officer: "The chain of command was obsessed with finding HVTs [high-value targets]. They received an intelligence report from 'another government agency' that said there was a possibility an HVT was on the other side of the mountain near Mana. The operations officer wanted the platoon to split so he could report to higher headquarters that he had part of the platoon in position near Mana as quickly as possible without being slowed down by the broken Humvee. I don't think he thought anything bad could happen; otherwise, he never would have told the company commander to split the platoon in front of everyone."

Note: What exactly the operations officer said to the company commander we'll never know for sure. Yet, as any subordinate can attest, the words a higher-ranking individual uses to issue an order rarely tell the full story of what those words imply. History, tone of voice, and body language all combine to create the context of the words' actual meaning. **An important takeaway for future leaders is that the moment you choose to go along with, pass** on, or turn a blind eye to an order that doesn't make sense, you become complicit in the senselessness. The law of gravity will always ensure that shit rolls downhill. One of the ways common sense leaders take care of their people is by standing up to and/ or standing their ground to block or deflect it.

Battalion Commander: "So the platoon had been dragging around this Humvee that had been broken. It was an up-armored Humvee. I think it was three days in a row, everywhere they went through Afghanistan. I'm sure you've been there, Sir. It's a pretty difficult maneuver to carry that thing around. So they had called and asked for an air assault to lift it out. So we went back to the aviators. The things were like twelve thousand to thirteen thousand pounds, which was heavier than the lift capacity of a Chinook helicopter we could get in there. So we told them, 'No. Keep dragging it.' We're going to work a way to do a ground evac for you. So they came to the vicinity of where Tillman got killed, and we worked a way to bring in some vehicles. And I say 'we,' it was really the CFT in Khowst, I'm trying to avoid pointing fingers at anybody. It was not anybody's particular fault. So they told the platoon leader to 'divide your platoon in half, we want you to take the HUMVEE and meet up with the wrecker, and we'll ground evacuate that so you can get on with your job without dragging that thing around.'"

Note: The empty weight of this particular Humvee was 6,500[54] pounds. The sling load capacity of CH-47D helicopter is 26,000 pounds, though the maximum capacity is affected by temperature, winds, fuel loads, and altitude.

Investigator: "Who gave that order?"

Battalion Commander: "The operations officer set up the logistical effort, but the company commander was with him. The company commander."

Investigator: "Was the operations officer aware that they were dealing with the issue of the split?"

Battalion Commander: "Oh yes, Sir…they all had situational awareness. They could hear, you know, the talking. To be honest, Sir, I don't think that it was a bad decision. We had been, we had been all through that area as a battalion in very small convoys. **The results that caused Corporal Tillman's death really had nothing to do with splitting that platoon up, except for that the converging forces killed him.** In my mind, the real problems occurred after that. I would be very comfortable, even now, sending out four or five vehicles just about any place in Afghanistan to fight. We do it all the time, and I don't think that was a bad decision."

Investigator: "Who specifically made the decision to split the platoon?"

Battalion Commander: "Sir, I believe that the operations officer told the company commander to split, and here's why: because I, we had pushed this mission off well, in general, they had been slipping the timeline because of these vehicles. I think that the company commander was back here planning, and we needed to get the whole company in to do this next attack. So they said, 'Hey, Sir, let us just take this up here, take the whole platoon up. We'll come back up, and we'll get over here.' And the operations officer was like, 'No, you'll never get it all done. Let's at least get that part of the platoon over here and get them all set.'"

Investigator: "Okay, why do you think the operations officer would indicate that he had no prior knowledge of the platoon split?"

Battalion Commander: "I don't know, Sir. To the degree he didn't know it, it would only have been because he wasn't listening. He sits six feet away from the company commanders in the CFT, so he can hear all their radio transmissions. I'm surprised that he would tell you that; his integrity is impeccable. If he says that, I believe him."

Since the XO was the recipient of all the verbal orders and directives from the chain of command, as well as the messenger that delivered them to the platoon leader, it's important to include his version of what he was told and the logic of why it was said.

Investigator: "When and how was the order issued to the platoon leader to conduct his movement on 22 APR 04?"

XO: "The FRAGO[55] was relayed over email. I issued the final order to move from the broken vehicle site over to Mana."

Investigator: "What did the FRAGO state?"

XO: "The FRAGO stated that the platoon leader was to move his element in two separate sections. One section was to move with the broken vehicle to the LU [link-up] point, and one section was to move into the next zone and prepare for future operations."

Investigator: "Who told the platoon leader to split his platoon into two sections?"

XO: "I received the order from the company commander, who received the order from the operations officer."

Investigator: "When the platoon leader expressed his concerns about splitting his platoon into two different sections, what was your commander's response?"

XO: "I agreed with the platoon leader's concerns about splitting his platoon into two separate sections. I relayed the concerns to the company commander, and he understood it was a valid concern, but **it was stressed by the operations officer to get forces in zone and not let the broken vehicle impede operations**."

Investigator: "Who told the platoon leader to move during daylight hours?"

XO: "I received the order from the company commander, who received the order from the battalion operations officer."

Investigator: "Are you aware of the battalion commander's directive that states no daylight movements unless approved by him? Any deviation from his guidance would need his approval?"

XO: "No, I didn't know that there's a direct prohibition against them. It was my understanding that the battalion commander strongly discouraged them and they were to be avoided whenever possible."

I asked Sergeant Major (ret.) Ted Kennedy, who served as A Company first sergeant from 2001–2003, what he made of these conflicting comments: "My experience is that many leaders want detailed descriptions of issues so they can make the decision for the leader on the ground versus listening to the recommendation made by that leader. If the XO supported his old platoon's recommendation to stay together, why didn't he advocate for it? Why didn't the company commander support the recommendation, and if he didn't agree, why didn't he discuss it with the platoon leader? Lack of trust, as well as fear of toxic leaders, is the obvious issue here! In a toxic climate with disliked leaders, the 'logic of why' is not something junior leaders think too much about...they know they'll be accused of insubordination if they ask why."

All four investigations confirmed that there was no actionable intelligence or general intelligence of any kind that made the movement to Mana that night, or any night, imperative. The second and third investigations summed it up this way: **"The comment by the operations officer/S-3 that 'we had better not have any more delays due to this disabled vehicle' contributed to an artificial sense of urgency to get boots on the ground before nightfall within his subordinates."**

Our purpose paves our choice of paths. For our choice of paths to make sense our purpose must make sense. As the Rangers of 2nd Platoon were about to learn firsthand, when our purpose doesn't make sense ("split the platoon to get boots on the ground before nightfall"), the paths we choose to accomplish it can't and won't ever make sense.

Magara, 5:30 P.M.: Although the sense of urgency to split the platoon and get boots on the ground in Mana didn't make sense to the Rangers of 2nd Platoon, getting out of Magara did. According to one of the Ranger sergeants: "There was a lot of confusion about the guidance from higher and the timeline. Every time someone would explain it, someone else would say, 'That can't be right.' The hardest part was sorting out who was going back and who was going on. We had to cross-load weapons, equipment, and personnel accordingly."

The downstream effect of splitting the platoon and ordering specific individuals within the platoon to go back to the base in Khowst was significant. In order to properly cross-load the remaining personnel, weapons, and equipment, the platoon's organic "chain of supervision" had to be broken apart. Individual Rangers such as the Tillman brothers found themselves reporting to people with whom they hadn't normally worked.

Breaking apart organic teams creates an impediment to effectively executing the tactics, techniques, and procedures that small units spend so much of their time training together and refining. Equally important to knowledge of tactics, techniques, and procedures is the knowledge of knowledge. Over time, leaders learn knowledge of their individual team members strengths and weaknesses with regards to tactics, techniques, and procedures. **Knowing your people's strengths and weaknesses is foundational to making sense of what they are capable of and what they may need assistance with. This knowledge of knowledge relationship goes both ways.**

The consequences of the choices made by the Ranger chain of command while sipping coffee inside their secure CFTs were now spilling over onto the individual Rangers of the 2nd Platoon. As is so often the case, the lowest ranking Rangers were the ones getting burned. Not only was the organic task organization of the squads now compromised, so too was the manning of the crew-served weapons. As the vehicles readied to move, Kevin Tillman found himself standing behind the Mk 19[56] on the rear Humvee, despite never having manned a vehicle-mounted Mk 19 or, according to what he told investigators, "never having fired a Mk 19 before."

While Kevin Tillman found himself responsible for a weapon system he had never fired in Serial #2, Pat Tillman found himself separated from his organic squad leader and sitting on top of a bunch of MRE boxes in the back of a Toyota Hilux pickup truck in Serial #1. Pat Tillman would now be working for and with a new squad leader while operating alongside new squad mates, some of whom were also separated from their organic squad leaders. **Although not a fatal blow to the platoon's chain of supervision (every unit trains for this contingency by reinforcing common tactics, techniques, and procedures), the ad hoc task organization added one more subtle constraint to the platoon's overall ability to adapt to whatever chaos the future might throw their way.**

As the Rangers continued cross-loading the vehicles and conducting final preparations for movement, the platoon sergeant and some of his squad leaders huddled together to study the map. The task that Serial #2 was given was to "escort the Jinga truck and inoperable Humvee on a route from their current location in Magara through a town called Tit and continue due north to their link-up location on the Khowst-Gardez highway." (See map.)

Eight days earlier (14 April), the platoon had traversed the same route (in the opposite direction) while driving from the forward operating base in Khowst to their final destination near Border Control Point #5. On the map, it appears as the most direct route. Their firsthand knowledge of the uneven, rocky terrain and steep grades made them realize that the drive would necessitate extreme caution and limit their already slow towing-speed to a snail's pace. The driver of the Jinga, on whom they now had to depend, was both untrusted as a driver and untested as an ally, and for those reasons they did not run the specifics of the route by him while they were still in Magara.

Of all the reasons splitting the platoon didn't make sense, the most foundational was that it impeded, instead of enhanced, their potential to accomplish the purpose that their chain of command was forcing them to undertake. There was no way (day or night) that any Jinga truck in Afghanistan could physically drag and deliver the inoperable three-ton Humvee up and over the terrain that comprised the northern route from Magara thru Tit to the K-G highway.

Additionally, once the sun went down, visibility would drop to near zero. To steer clear of hazards such as crevasses and cliffs, you have to actually see the hazard first. Not an insurmountable issue for the Rangers, who were all equipped with state-of-the-art night vision devices and infrared headlights. However, Afghan Jinga truck drivers don't come equipped with night vision goggles. Which meant that the Jinga truck would have to use his headlights, and those headlights would then "bleach out" the infrared light used by the Rangers' night vision goggles and render them useless. This meant that most of the convoy would be forced to turn on their white lights to ensure that they didn't drive off the side of a cliff either.

No matter how hard the platoon sergeant and his squad leaders tried to make sense of it, there was no way to call this a tactical movement. Their angst wasn't based on a theory or a hunch but rather on firsthand knowledge that the entire platoon had already learned and shared.

The more they discussed the task they had just been given, the more the Rangers of 2nd Platoon realized the reality of their predicament and the dangers that now lay ahead. What started as anger and frustration with their chain of command had now transitioned to dread:

"This mission was one more check-the-box op that we were doing for no real reason."

"When you're doing something really stupid, and everyone knows it's stupid it's really hard to stay focused and motivated. I just kept telling myself, 'We're in bad guy land, and we could get ambushed at any minute.'"

"We didn't even have time to look at a map before getting back on the road. We were rushed to conduct an operation that had such flaws, which in the end would prove to be fatal."

To better understand and learn from the sequence of events about to unfold, it's helpful to put ourselves in the boots of the Rangers as they prepare to depart Magara. Many had only been in Afghanistan for twelve days, and those that had been on previous deployments had never worked in this area of the country. They had been living out of the back of their vehicles for eight days, during which time they had been rained on and endured freezing temperatures at night. They were short on food, water, and sleep. Based on the feedback gathered by investigators during interviews, they were totally frustrated with their predicament and with their chain of command.

Our ability to influence what's going on around us requires conscious awareness of what's going on *inside us* first. The term "stress" is defined as "the brain's response to situations involving conflict, danger, or demand for change."[57] Whether it's a life-or-death situation on the battlefield or an unexpected challenge to your business, conscious awareness of the causes and effects of stress, as well as the simple, time-tested techniques to instantly diffuse or counter stress, are something every leader should know. To survive any type of situation, no matter how simple or complex, **follow the Delphic maxim, "Know thyself."**

To Be Continued.

CHAPTER 7

Five Facts About Emotional Stress and How to Counter It

Fact #1: Stress is not caused by the situation going on around us; it's **caused by the way our brains perceive and respond to the situation.** As an example, the stress response of fear is triggered when our brains perceive that we're in danger, whether or not the danger is real or imaginary. (e.g., "Is it a snake or a hose?) All mammals respond to stressful situations, such as the sudden appearance of a snake on the path or a senseless order from a toxic boss, by releasing hormones including adrenalin and cortisol. Cortisol is the first stage of flight or fight. It mobilizes glucose to the thigh muscles and increases our heart rate, enabling our bodies to run faster and/or fight harder.

In the short term, adrenalin and cortisol are brilliantly adapted to help us survive unexpected physical threats, such as carrying a wounded teammate to safety or getting out of a building to escape a fire.[58] But non-life-threatening contemporary stressors, such as those we experience at work, in social settings, or while worrying about the future, can also trigger the release of cortisol and its behavior-altering effects. Studies show that short-term exposure to stress-induced cortisol inhibits brain function, specifically the neurons in the brain relating to learning, memory, and judgment.[59] *"I was trying to subtract the distance we walked after the previous turn, but I was so pissed off I couldn't do the math."*

Situations that create long-term exposure to cortisol—such as working under a toxic leadership climate—have been shown to inhibit the release of oxytocin, the neurochemical responsible for empathy and compassion. Without empathy and compassion, people stop helping each other. Long-term exposure can permanently destroy brain cells and cause chronic conditions such as heart disease, diabetes, and PTSD.

The way our individual brains manage the cortisol stress response could explain why some veterans suffer from PTSD while others who served by their sides do not.[60] We don't know enough at this time to say for sure as more testing is needed, but we do know that cortisol is a killer and toxic leaders are cortisol creators.

Fact # 2: Emotions underlie the majority of the stress we experience. We call emotions "feelings" because they are physically present in our bodies and brains as neurochemicals that we can feel. The human brain evolved to categorize incoming information in terms of emotions, including fear, sadness, anger, pride, happiness, and compassion.

All emotions do the same thing in different ways. They encourage us toward things associated with pleasure (eating, drinking, sex, social status, etc.) and steer us away from potential pain (embarrassment, starvation, things we've never experienced before, etc.). By tabbing memories according to their potential for pleasure and pain, our emotions enable our brains to learn from our experiences and to rapidly focus on those that are context-specific.

Herein lies the catch. In order to induce a focused response, our brains evolved to inhibit knowledge that doesn't match our current emotion. Thus, when the emotion of fear is present, our brains suppress access to knowledge of previous acts of bravery and false alarms. Anger blinds our brains to the knowledge of compassion and calm. "Blind rage," "blind lust," and "blind ambition" are all byproducts of our emotional brain's capacity to inhibit or blind our brains to alternate options. This creates defective single-track strategies, which almost always yield flawed results.

Fact #3: Emotional stress causes cortical inhibition. The problem with our emotions is not that they exist—we need emotions to learn from our experiences—**the problem is our difficulty turning them off** *before* they initiate what brain researchers refer to as "cortical inhibition."[61] Cortical inhibition is the blocking of actions stemming from the neocortex (aka our thinking brain). The symptoms include but are not limited to the following[62]:

- Diminished ability to think clearly.
- Less efficiency in decision-making.
- Diminished ability to communicate clearly.
- Reduced physical coordination.

The neocortex is the only part of our brain that is able to consciously monitor and make sense of sensory information. Thus, cortical inhibition can limit everything from what we see, hear, and smell to the way a person thinks and feels, to how they move. If our neocortex can't make sense, we can't make sensible choices about what to do next. The phenomenon of cortical inhibition helps explain why sensible people choose to do senseless things.

"The entire platoon is dangerously cold, tired, frustrated, and red-hot pissed off. Which is why everybody—including you, Sir—are starting to do some really stupid shit."

Fact #4: The first step toward reducing cortical inhibition is learning what your triggers are. What follows are three of the most common:

a. Loss of freedom to control our own destiny. One of the most common causes of stress in group settings occurs when our brains perceive we've lost our freedom to control and/or have a say in our own destiny.[63] When trapped, denied sensory information, and/or forced to do something that doesn't make sense, we instantly recognize that we have no options. Without options, we have no freedom to choose between them and, by proxy, no freedom to control our own destiny.

To understand how foundational our sense of freedom to control our destiny is, we only need to think of situations where it's taken away. In the extreme: physically trapped, forcefully confined, or buried alive in a box. Claustrophobia is our nervous system's primordial reaction to the perceived and/or actual loss of freedom.[64] While it can be a life-altering disease for some, all humans experience claustrophobic fear to some degree when we are denied sensory information, perceive we are trapped, or are forced to do something that doesn't make sense.

In combat or business contexts, we experience the same reaction when we and/or our teams are ordered by toxic leaders to do something that doesn't make sense and then denied our freedom to speak up, out, or with them about it. What follows are common situational triggers that cause our brains to perceive we're losing our freedom of choice to control our own destiny:[65]

- The absence of helpful or predictive information such as: what you'll be doing next, or how difficult a challenge will be, or how long something is expected to take.
- A feeling that the situation going on around us is worsening and we are unable to do anything about it.
- A lack of social and emotional support.
- Feeling as if you're trapped and there's no way out.

b. Stress levels are influenced by our position in the hierarchy. It's not the title or job description (e.g., CEO, commander, etc.) that causes the most stress; rather, it's the degree of control workers feel they have throughout their day. As we've already learned, less control equates to more stress.

Leaders have far less stress because they are in control of the people who work for them, their direction, and what they will or won't do next. A forty-year-long stress study of 18,000 people in the British Civil Service confirmed that the risk of stress-related illness decreases as a person's rank in the Civil Service goes up.[66] And recent studies have confirmed that the lower the military rank, the higher the risk of developing PTSD.[67] Whether in combat, business, or life, **those who are most vulnerable to stress and its sense-inhibiting effects are the lowest-ranking members**.

c. Stress levels are influenced by the people around us. The capacity to transfer emotions between humans is innate. Emotional contagion is the tendency for humans to feel the same emotions as the people around them. Mirror neurons in our brains are tuned to reflect the emotion-driven behaviors of those around us (e.g., facial expressions, dilated pupils, panicky body movements, tone of voice, etc.) before we have time to consciously think about them.

The evolutionary basis of mirror neurons can be seen when a herd of deer or flock of birds detects a potential threat. One animal sounds the alarm by taking flight, and all others instinctively follow suit. It works the same way for humans. In life-or-death survival situations such as combat and natural disasters, the contagion of fear and panic spreads like wildfire, as do courage and calm. Don't forget that our emotional brain can be duped by its perception of what others are thinking or feeling. Only our neo-cortex/thinking brain is capable of pressure testing what we perceive against the reality of the situation as it's going on around us. *"He's really amped up, and he's still shooting, even though the enemy fire is ineffective."*

Fact #5: You control your mind, not the other way around. Arm yourself. Before going into life-or-death situations, first responders are taught to check their weapons, ammo, radios, and night vision goggles. **The Common Sense Way recommends they also check their neocortex to ensure they know how to use it to overpower their emotions, stay calm in a crisis, make better decisions, and think better thoughts.**

Remember the old saying that thoughts control our behaviors? It's been accepted as fact for centuries until a few years ago, when science discovered it actually works the other way around. Since then, a flood of studies have confirmed that we can control our thoughts and feelings through conscious control of our behaviors. What does this mean in practice? It means we can change the way we think and the way we feel by consciously changing our behaviors, and we can do it at any time, in any place, and under any circumstances. **All we have to do is engage our neocortex.**

Engaging our neocortex is essentially a technique where we pay attention to the present. By focusing our senses on the sights, sounds, and physical sensations going on around us in the here and now, we can achieve mental clarity about what's actually happening in our lives.

The Problem: Although the neocortex is three times the size of our older unconscious brains, its position on top makes it the last to receive bottom-up sensory information and the slowest to react. Studies show that when our physical senses sense something—such as the sight of someone moving on the ridgeline above us, or the sound of something going bump in the night—our reptilian and mammalian/emotional brains are alerted within twenty milliseconds, while awareness of the neocortex occurs 280 milliseconds (about ¼ of a second) later.[68]

As a result of this ¼-second "awareness gap," our brain's first instinctive response to any type of situation is always unconscious, emotional, and without context. We feel fear before we know what we're afraid of. We become angry before we realize what made us mad. And we make choices based on emotion-generated response options that were passed down to us from our ancestors instead of the adaptive stimulus of what's currently going on around us.

A common, contemporary example of the way our three-part brain perceives and responds occurs whenever another car cuts us off on the highway. Your brain's initial unconscious reptilian reaction is survival threat and violation of personal space ("he could have killed me, and I had the right of way"), amplified by the emotions of fear, anger, and aggression (aka road rage). Your follow-on, consciously-aware, neocortex-generated response option (¼ of a second later) might be the realization that the other driver is old or new, or driving a loved one to the hospital, or might not have actually seen you.

What do close calls on the highway and getting caught in an ambush have in common? **Our survival is on the line, and the most important choice we can make is to engage our neocortex—stay calm and think.** What follows are three time-tested, scientifically-proven techniques that anyone can use in survival situations to instantly engage their neocortex and vanquish anger, fear, and panic:

1. **Breathe Deep and Count.** Deep breathing while counting to yourself is the most simple and effective method of engaging your neocortex and instantly decreasing stress, anxiety, and high blood pressure. Which is why it's the core component of yoga and Zen meditation and the time-tested adage of taking "ten deep breaths" to calm down.
   How to do it: Deep breathing—also called diaphragmatic or belly breathing—takes place when you breathe so that your lower belly expands and contracts. When you inhale, in this

Brain evolution

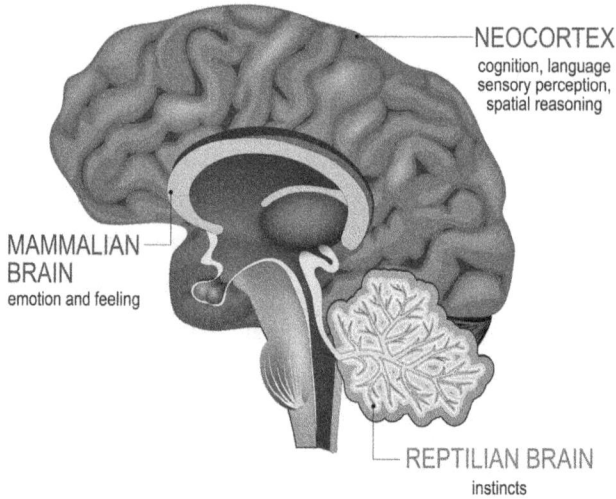

NEOCORTEX
cognition, language
sensory perception,
spatial reasoning

MAMMALIAN
BRAIN
emotion and feeling

REPTILIAN BRAIN
instincts

manner, your diaphragm moves downward and pulls your lungs along with it. When you breathe out, your diaphragm pushes upward, helping your lungs expel carbon dioxide. Give the 4-4-8 technique a try right now. Take a deep breath through your nose while counting up to four, now hold your breath for five, six, and seven, then exhale through your mouth while counting backwards from eight. It helps to place your hand on your belly while breathing to feel the diaphragm expand and contract. Repeat five to ten times.

2. **Say What You're Thinking and Feeling Out Loud.** Giving voice to what we're experiencing can change our perception of it. We can't fully illuminate our thoughts and feelings unless we translate them into something tangible. To accomplish this, our neocortex interprets our thoughts and feelings through the use of language. Words matter; they're how we crystallize and convey our thoughts. Calmly saying what we think or feel out loud is an evolutionary

sense-making ability that enables us to collaborate with our other senses, as well as those of the people around us, to see if what we're thinking actually makes sense.

How to do it: Continue deep breathing while talking to yourself and to those around you in a cool, calm, and collected manner. It's the secret defense against impulsiveness and psychological paralysis in life-or-death survival situations. Actions control thoughts. Take action and control your emotional thoughts by calmly saying what you're thinking out loud. **When we act calmly, we calm the way we act.**

3. **Pay Attention Like a Cat**. To understand what's going on around you—whether you're a sniper on a reconnaissance mission or a parent driving the family home on a dark, stormy night—focus your eyes, ears, nose, mouth, and brain like a cat searching for mice in a field.

   How to do it: Give it a try next time you're driving at night. Breathe deeply and consciously focus your eyes, ears, nose, and brain on the road in front of you. Behold how your senses sharpen. You will see clearer, hear better, and have more sensitive feel and control of the steering wheel, accelerator, and brake.

Why These Techniques Work:

The neocortex is the only part of the brain that can take conscious control of an unconscious process such as breathing. The neocortex is also the only part of our brains that can count and create speech. Another way to look at it is that it's impossible to consciously control our breathing and/or speak calmly without engaging our neocortex.

One of the most important principles to understand about the neocortex is that it only engages when you pay attention to what's going on around you. In order to pay attention, our neocortex needs a purpose (e.g., to breathe, eat, drink, escape danger, seek companionship, etc.). "Paying attention like a cat" provides our neocortex with the logic of purpose. Sensory information (sights, sounds, smells, taste, and touch) provides the common language that our nervous system uses to recognize patterns, make sense of what's going on around us, and make sensible choices about what to do next. As we consciously pay attention by focusing our senses on the situation around us, our neocortex engages, our senses sharpen, and we can make better decisions and think better thoughts. **Believe in and trust your senses.**

CHAPTER 8

The Firefight

6:00 P.M., Driving out of Magara: Once the platoon sergeant and squad leaders had everyone loaded in the proper vehicles and the vehicles lined up in the proper serials, they conducted one final communications check to confirm all radios were working properly. Satisfied that they had positive communications between all vehicles, the six vehicles of Serial #1 drove out of Magara on their way to Mana.

6:10 P.M.: Less than one kilometer north of Magara, Serial #1 arrived at the Y-intersection (see map below) and turned left

Serial #1 (16 Rangers, 7 Afghan Mil)

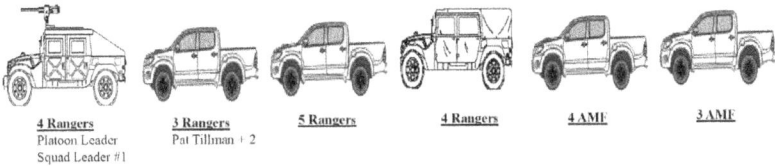

4 Rangers	3 Rangers	5 Rangers	4 Rangers	4 AMF	3 AMF
Platoon Leader	Pat Tillman + 2				
Squad Leader #1					

Serial #2 (19 Rangers, 2 Afghan Civ)

2 Rangers	6 Rangers	4 Rangers	4 Rangers	3 Rangers
Afghan Civilian Drvr	Squad Leader #2	2 x Mortar Section	Regimental SGM	Platoon Sergeant
	Drvr	2 x Snipers		Kevin Tillman
	50 Cal gunner			
	M240B gunner			
	M249 Saw gunner			
	M203 gunner			

163

Serial #1 arrives Y-intersection while Serial #2 prepares to move out of Magara. Vehicle size is not to scale with map. Distance from Magara to Y intersection = 900 meters; from Y intersection to entrance to slot canyon = 1.4 km; from slot canyon entrance to spur = 1 km. Total distance from Magara to spur = 3.3 km.

onto Canyon Road, which would take them directly to the town of Mana.

A few minutes after taking the turn, Serial #1 entered the mouth of a dramatically narrow slot canyon[69]. Its towering rock walls and horseshoe-shaped path were carved into the mountain over millions of years by the same waterway that was now splashing beneath their Humvees. The close-quarter walls kept the sunlight out and the temperature low, which prompted some of the Rangers to don their Gore-Tex jackets over their uniforms. In some spots, the passage between the walls was only a foot or two wider than the width of their Humvees. The going was slow, and the margin for error was mere inches.

During interviews with investigators, many of the Rangers described the canyon as both a funnel and a maze. Another way to think of it was as an alley. Envision driving through an alley bordered by enemy-occupied buildings, and you can imagine how vulnerable the Rangers felt as they passed through the canyon. With heads and necks straining to keep a sharp lookout skyward, every Ranger in the convoy was on high alert. "I've got a bad

Entrance to slot canyon.

feeling about this," was a common refrain. Seven minutes after they entered, Serial #1 exited the canyon without incident.

After exiting, they continued following the creek bed west for another five hundred meters to a point where it forked just north of the town of Mana. Unsure of their exact location, they pulled over to orient themselves on their maps. Directly above them were five adobe-like homes. An old man and four to five children stood outside and stared. After a few seconds, the children ran inside.

6:05–6:10 P.M.: Five to ten minutes after Serial #1 had departed, the six vehicles of Serial #2 drove out of Magara toward the Y-intersection. The going was much slower for Serial #2 due to the Jinga truck and towed Humvee moving in the middle of the convoy (third in order of movement). The Rangers of Serial #2 instantly realized that having the Jinga truck and towed Humvee embedded in their convoy was more than just an inconvenience and added irritant. **No matter where the Jinga truck was positioned in the convoy (e.g., front, back, middle), it severely impeded Serial #2's best defense against any type of enemy attack—their freedom to fire and maneuver as a team.**

After the lead Humvee in Serial #2 arrived at the Y-intersection, they did what the plan directed them to do. They took a right toward the town of Tit, as did the Humvee behind them. When the Afghan Jinga truck driver, who wasn't privy[70] to the plan, arrived at the Y-intersection, he instantly recognized the error of their way and slammed on his brakes to ensure they'd listen to what he had to say.

The Rangers' interpreter asked the driver why he was stopping, and the driver immediately explained what everyone already knew: that "the route through Tit wasn't possible" and "there was no way that he and his Jinga truck would ever even consider trying it." He further explained to the interpreter that "the fastest and safest way to get the vehicle to the K-G highway was to take a

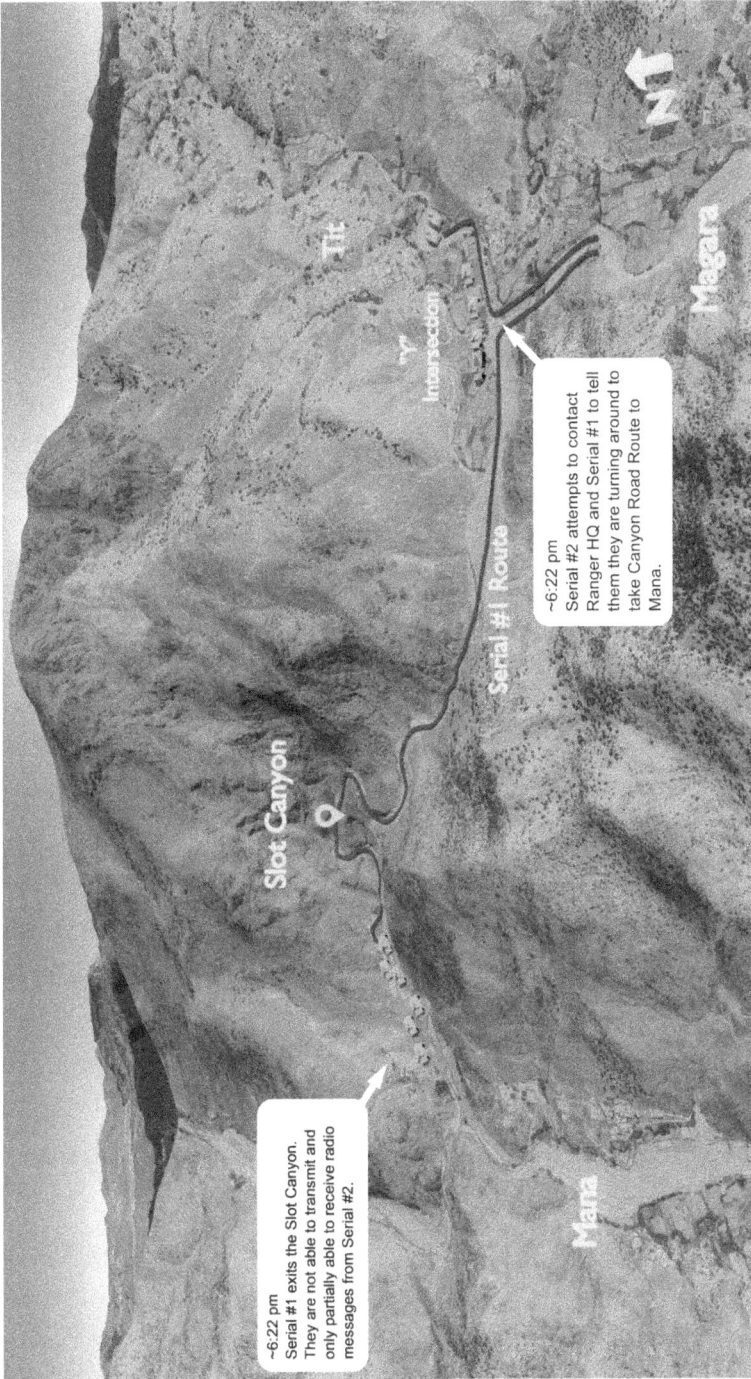

Serial #2 arrives at Y-intersection and discovers their planned route is not feasible.

left instead of a right at the Y-intersection and follow the Canyon Road," which was the same route that Serial #1 had just driven.

With the convoy at a standstill, the platoon sergeant radioed the other vehicles to halt, then jumped out of his Humvee so he could see what the Jinga truck driver had to say. As he later explained: "We had driven that route down before, and it's really tight in some areas, and with the Jinga truck and the towed Humvee, my main concern was we were going to go off the side of the hill at night. The Jinga also had to move white light."

When the platoon sergeant was asked by the investigators, "Did you contact anyone to let them know that you were changing route?" The platoon sergeant explained: "So I attempted at that point to try to get a hold of the platoon leader, but I think due to the antenna—we had a vehicle-mounted antenna—we couldn't get COMMs."

The fact that the platoon sergeant's UHF radio did not allow him to communicate effectively with the platoon leader was as understandable as it was predictable. UHF radios cannot communicate with each other unless the two antennas have "line of sight" between them. "Line of sight" means a direct line between the antennas without intervening obstructions. Thus, line-of-sight radios don't communicate well when their antennas have a mountain between them. This is a fact known by all infantrymen and was one of three key points made by the platoon leader when he explained the logic of why splitting the platoon didn't make sense. We know that the platoon sergeant's UHF radio was working properly because he tested it before departing Magara, and he had just used it to tell the rest of Serial #2 to halt before he got out of his vehicle.

During the fourth and final investigation (2005–2007), the platoon leader was asked for the first time about whether he received a radio transmission from his platoon sergeant while he was driving with Serial #1 toward Mana: "I could hear him attempting to contact me. Eventually, I could not reach him on my radio, but my squad leader [who was driving and sitting next to him] got in touch with him on his radio." This squad leader, from hereafter referred to as "Squad Leader #1," told investigators: "I heard over the radio that Serial #2 was going to change routes because the local national driver towing the broken-down Humvee said the route was too dangerous. I asked the platoon leader if he heard the transmission, and he said yes. I believe both of us attempted to get into contact with Serial #2 to tell the platoon sergeant that we knew they were following behind us, but to the best of my knowledge they never heard us."

The platoon leader added: "The Jinga driver was protesting the route we planned for Serial #2, citing 'it was going to be too steep for the Jinga truck to tow the Humvee, and he did not want to risk it.' **I was able to say, 'OK, if there was another route to get to the same place, go ahead and go with it.'** During communication, it was real spotty and breaking up. Therefore, I would say the communication was broken at best, but not effective."

The UHF radio signal was just strong enough for the platoon leader and Squad Leader #1 to hear that Serial #2 had abandoned the planned northern route through Tit. Tragically, the radio signal was too weak for them to discuss and coordinate what Serial #2 would do next. While Squad Leader #1 told investigators he understood Serial #2 would be following them on the same route, the platoon leader said he did not. There is no record of either the platoon leader or Squad Leader #1 sharing what they heard on the radio with the rest of Serial #1.

When we think of the negative effects of stress-induced cortical inhibition, we usually think of how it inhibits our capacity to recognize and respond to threats. Yet equally debilitating and dangerous is how it inhibits our brain's capacity to recognize options as opportunities to avoid those same threats. Studies show that when we are under stress, we pay less attention to changes in the environment, which reduces our potential to recognize options as opportunities to avoid or counter a threat.

Even though the platoon leader was aware that Serial #2 was taking an unplanned route, the false sense of urgency created by their purpose—to get boots on the ground in Mana before night-fall—prevented him from slowing down or stopping to look at his map and figure out what other routes were available. It would have taken mere moments of studying the map to recognize there were only two routes from Magara to the K-G highway (see map, page 164). The first option was the northern "Camel Trail" route through Tit that the platoon had struggled to traverse eight days earlier and that the Jinga driver had confirmed as impassable for a truck towing a Humvee.

The only other option was the Canyon Road to Mana that Serial #1 was currently driving. This route followed a flat, dry creek bed all the way past Mana to the K-G Highway. As the Jinga truck driver told Serial #2, the Canyon Road route was "far faster and safer" than the Camel Trail route that their chain of command had implied they should take by telling them to split the platoon.

Undeterred by his inability to make "positive comms" with his platoon leader, the platoon sergeant did what all leaders who find themselves in similar circumstances are supposed to do: "I started trying to call on the satellite radio to the TOC, but I couldn't get a hold of anybody. I attempted to talk with the TOC and with Serial #1 to advise our plan of action to route behind Serial #1 and head north on another route, but met with negative comms over the radio."

Unlike UHF radios that require a line of sight to communi-cate, a satellite radio bounces its signals off of satellites in space, which makes it highly reliable in almost any type of terrain or weather condition. So why wasn't the platoon sergeant able to establish satellite communication and receive assistance at this critical moment from his battalion headquarters in Khowst or the Ranger regimental headquarters in Bagram?

There are some common problems that are known to prevent satellite radios from communicating, such as calling on the wrong frequency, the satellite antenna not being oriented properly, or getting "stepped on" by other communications traffic, etc. Yet all of these problems are both easily correctable and so common that the immediate action steps to fix them are second nature to anyone who regularly uses a satellite radio. Whether any of these problems were present that day, we will never know. However, based on what we do know from the statements and evidence collected and recorded by the investigators, there is reason to consider an alternative explanation for why the platoon sergeant wasn't able to establish satellite communication and receive assistance at this critical moment from the Rangers higher headquarters.

Contained within the "first draft" of the first investigation under the heading "Findings" is a sentence written by the investigator concerning command and control. Paragraph B.4 states the following: **"There is evidence that the battle staff was not actively tracking the platoon's movement and was unaware the platoon initiated movement during daylight hours."** After this sentence, someone who reviewed the report inserted the comment, *"If it's not important to the events that led up to actions before/during/ after the ambush, then leave it out."* The sentence was left out of the final/official version of the first investigation and is not present in any of the other three investigations.

The importance of the sentence concerning what the reviewer described as "the events that led up to actions before/during/ after the ambush," is self-evident. **One of the highest priority categories of mission-critical information is any change to the**

planned route. Especially changes that involve the potential for friendly forces to converge on the battlefield.

There is always a possibility that the satellite radios used by the Ranger chain of command at both the forward operating base in Khowst and at Bagram weren't working properly at the exact moment the platoon sergeant tried to contact them. However, even if the satellite radio wasn't working properly, the Ranger chain of command would have still been able to see Serial # 2 move out of Magara via the blue force trackers.

Additionally, by monitoring the blue force trackers, they would have also been able to see Serial #2 turn around and begin moving on a collision course with Serial #1 as well as another, previously unknown, Ranger platoon that was heading for the exact same spot.

The 2nd Ranger Battalion Reconnaissance Platoon, with eighteen Rangers riding in three Toyota Hiluxes, had been operating in the mountains for six weeks. The three Toyota Hiluxes were equipped with three satellite radios and three UHF radios that were all in use and working properly. A few hours earlier that day, the operations officer/S-3 had contacted the reconnaissance

Blue force tracking systems use satellite beacons hard-wired into each vehicle to enable viewers to see the exact map location of all blue/friendly force vehicles. The view in the picture above is from a hand-held blue force tracker.

platoon over the radio and ordered them to return to the forward operating base in Khowst so they could refit and prepare for the next mission. To get back to Khowst, the reconnaissance platoon had to get to the K-G highway. The only way for them to get to the K-G highway was to pass through Mana.

The Ranger reconnaissance platoon was not mentioned in a single one of the investigations. The discovery of the reconnaissance platoon's existence, location, and lack of situational awareness happened in January 2023, nineteen years after the firefight. I had just finished talking with one of the Rangers from 2nd Platoon and asked him if "there was anything else about the firefight that might be important?" He thought about it for a few seconds and then said, "When I got to the spur, one of the first guys I ran into was the reconnaissance platoon sergeant, and we both looked at each other and said, 'What the fuck are you doing here?'"

I asked him if he remembered the name of the reconnaissance platoon sergeant, he replied, **"Of course, his name is [Sergeant Rex]; do you want his contact information?"**

Fortunately I knew Sergeant Rex from my last tour as a Ranger. He and I designed and built the 2nd Ranger Battalions first shoot and move live-fire range. He served in every leadership position from squad leader, to command sergeant major, and was/is highly respected throughout the Ranger Regiment. I was able to make contact with him the same day I got his contact information. Within minutes of listening to him talk about his experiences with the reconnaissance platoon, I realized how much key information he possessed about the events before, during, and after the firefight. When I asked him why none of this information was contained in any of the four investigations, here's what he said:

Sergeant Rex: "To this day, I still have no idea. I was one of the first Rangers on the scene after the firefight ended. At the time of the investigations, I couldn't believe they didn't want to interview me. And it wasn't just me; quite a few of the NCOs in the battalion asked me, 'Why aren't the investigators talking to you?'"

None of the four investigations interviewed a single member of the reconnaissance platoon. Incredibly, nineteen years after the firefight, only one member of the 2nd Platoon that I talked to knew that the reconnaissance platoon was anywhere near them at any time that day.

I asked Sergeant Rex, "Why do you think they didn't interview you?"

Sergeant Rex: "I'm not sure, but I've thought about it a lot. The only reason I can come up with was because the chain of command didn't tell them we were out there. They knew that if the investigators talked to us, we'd tell them that none of us knew where each other were at, and the only way that happens in a combat zone is if the commanders and their massive staffs aren't doing their jobs. They were so obsessed with finding HVTs that instead of doing what they were supposed to be doing, which is to provide command, control, and communications to the men in the field, their priority was planning the next dry-hole instead."

Of note, there is no record in any of the four investigations that any member of the chain of command was asked whether or not they were monitoring the satellite radio and/or the blue force trackers during the time window that all three forces were driving obliviously on a collision course near Mana.

What lesson can past, present, and future leaders learn from the Y-intersection near Tit? It's the lesson of purpose. The purpose of all leaders is to take care of the people they have been given the privilege to lead. Taking care of our people isn't some touchy-feely thing that means we coddle them. **Taking care of our people means making good decisions and solving complex problems that set the conditions for our people to succeed. That's how we create a healthy leadership climate.** The

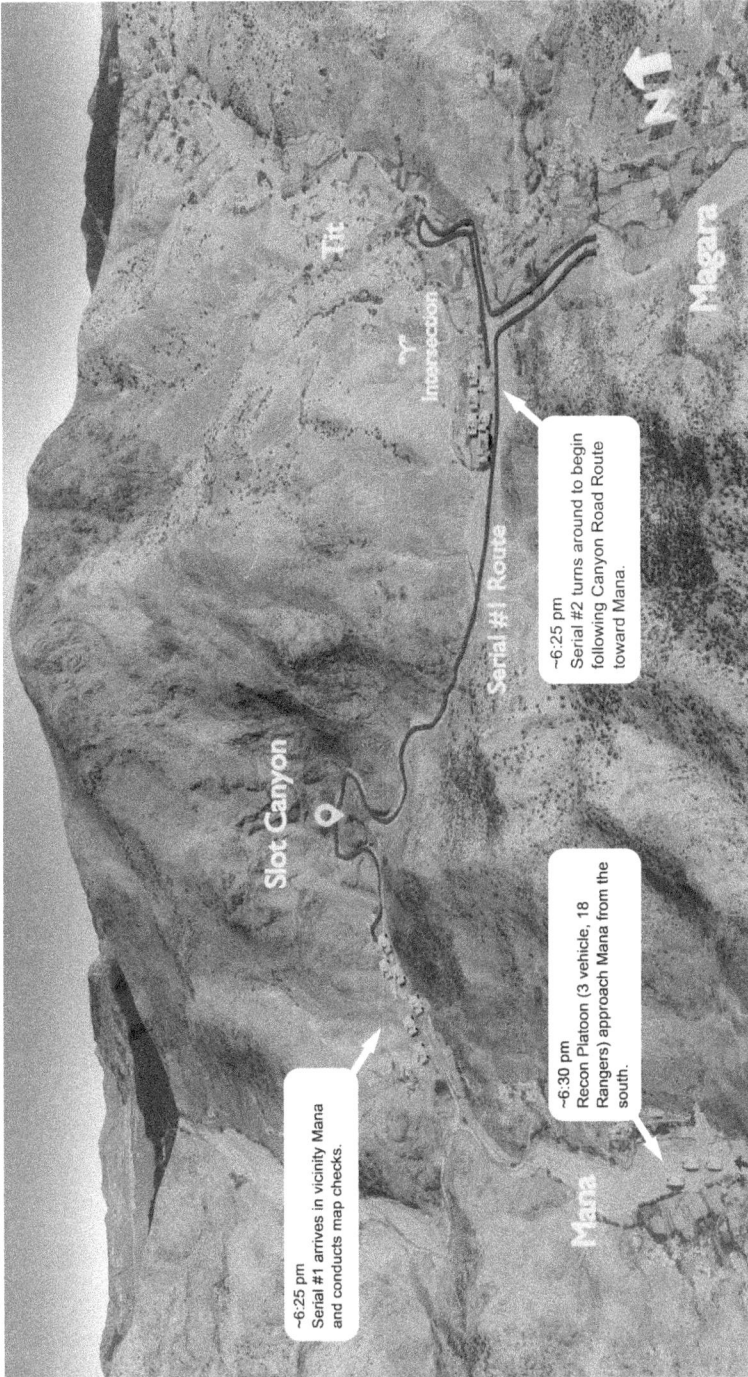

Disposition of three separate forces who have no communications between them and no knowledge of each other's locations.

Y-intersection provides a prime example of the logic of why this principle is so foundational.

If the platoon sergeant had been able to talk to any member of his higher headquarters chain of command, such as his company commander, the battalion operations officer/S-3, the battalion commander, or the regimental commander, and/or any of their deputies, any one of those leaders could have provided the three converging forces with a communication bridge to connect and pass mission-essential information between them.

Additionally, any one of those individual leaders and/or their staffs could have used their non-stressed brains to recognize that there was only one route option available for Serial #2 to tow the inoperable Humvee to the highway. After discovering this, **they would have instantly recognized and alerted Serial #1, Serial #2, and the reconnaissance platoon to the catastrophic potential unfolding on the ground as three armed-to-the-teeth forces, who were unaware of each other's locations, were about to converge.** Critically, they would have been able to prime the minds of the Rangers in all three forces with the life-saving guidance of *"keep a sharp lookout for the other serial and/or the other platoon, they are just up ahead or behind you."*

To ensure future leaders take away the common sense lesson of the Y-intersection, what follows is one version of how the communication bridge conversation could have taken place over satellite radio between the platoon sergeant and any member of his chain of command:

"This is Platoon Sergeant, Serial #2 is currently located at the Y-intersection outside Magara[71] and we've determined that the route north is impassable. The only way to tow the Humvee to the K-G Highway is to take the same route that Serial #1 took down Canyon Road past Mana. **Break.**[72] We are currently turning around to take that route but are unable to make positive comms with Serial #1 and need you to act as a communication bridge and relay for us. OVER."

"This is Commander, I copy the route north is impassable and you are turning around to follow the same route as Serial #1. If situation allows, stand-by at your current location until we are able to make positive comms with Serial #1 and the reconnaissance platoon to ensure they acknowledge the route change. OVER."

Platoon Sergeant: "Roger, we are able to stand by at the Y-intersection until we hear back from you that you have made positive comms with Serial #1 and the reconnaissance platoon and that they have acknowledged the change to our route. OVER."

Once a communication bridge was established between the three elements and all key information had been passed and acknowledged, it would be the responsibility of the higher headquarters leader and his staff to continue monitoring the situation and, whenever needed, continue providing a communications bridge between the three elements until the mission had ended.

6:25 P.M., Serial #2, Y-Intersection: When a convoy changes directions for any reason (e.g., gets lost, takes a wrong turn, or realizes the route is impassable, etc.), the leader of the convoy feels the frustration it causes the passengers as well as the perception—real or imaginary—that whatever caused the change is the leader's responsibility. Because the brain can only think of one thing at a time, the platoon sergeant's priority was now focused on ensuring that the new route recommended by the Jinga truck driver would not cause the convoy to turn around again. Recall that the platoon sergeant had never been down Canyon Road before and, like the other members of Serial #2, had neither the time nor the reason to study the route on the map before they departed.

Once again, the false sense of urgency created by the order to get "boots on the ground in Mana before nightfall" caused the leaders of Serial #2 to fall victim to one of the most common oversights made when convoys change directions: not immediately informing the rest of the convoy where they are heading and why.

It happens to families on vacation, it happens to co-workers on business trips, and unbeknownst to the men of Serial # 2, it was happening to them at that moment.

The third investigation summarized what happened next: "Communications between and within the serials was imperfect at best. Not all soldiers in the serials understood the movement plan for the two serials, particularly after Serial #2 changed its route. Accordingly, some soldiers lost situational awareness to the point they had no idea where they were or where they were in relation to the other serial."

The only thing most of the Rangers in Serial #2 knew about why they were turning around came when the platoon sergeant called the squad leaders on the radio and told them: "The Jinga truck driver knows a better way, and he's going to take the lead." One of the Rangers riding with Serial #2 told investigators that the on-the-fly radio transmission was not shared or explained to his vehicle. "I had no idea what route Serial #1 took, and I had no idea that by turning around we were following right behind them."

The common sense rule of thumb is this: whenever you receive updated information or learn of changes to what everyone on your team knows or believes to be true, you must share that updated information with them as soon as possible and without filtering it. The importance of this principle is to ensure that everyone is on the same sheet of music with regards to their purpose, so that in the event their leaders are incapacitated or incommunicado, they are able to carry on with the mission.[73]

If soldiers don't understand why they're doing what they're doing, they have no common sense of shared purpose to pivot from. Without a common sense of shared purpose, our individual brains all come up with our own: *We must be heading back to Magara; maybe we forgot something; maybe HQ is messing with us again, etc.* Our purpose paves our choice of paths. When our purpose is different, so too are the paths we choose to accomplish it. As a result, when the situation going on around us changes, we all make sense of what we're experiencing in different ways, and we all make different choices about what we should do next.

Sergeant Major (ret.) Ted Kennedy: "The cardinal rule must be ingrained in the culture of the unit. When higher is telling lower, 'just do as you're told and follow the order,' or 'you don't have a need to know,' or 'because I said so,' that culture rolls downhill. Why would a squad leader feel a need to tell a team leader or a Ranger private everything, when he himself is not being told anything? The culture of following orders blindly has always been part of the military. When officers and NCOs are afraid to ask 'why' that mentality rolls downhill. I know this because I was a part of that culture and I was also a party at times to saying, "because I said so." This is one of the saddest traits of the Rangers, and I spent seventeen years Rangering. I wish someone would have taught me and inculcated in me a culture of the 'logic of why' at a very young age."

6:30 p.m., Y-Intersection: Recall a leadership climate is created by the sum total of choices made over time by all leaders within the climate system. Over the previous twenty-four hours, the Rangers' chain of command had made and enforced the following choices and orders:

Don't leave your inoperable Humvee in a secure base (BCP-#5) where it can't cause you any more problems; instead, drag it with you through enemy-occupied mountains where it can.

Permission to blow up the vehicle is denied, we don't care what happens to it, just tow the vehicle along with you until you can't.

Get boots on the ground in Mana before nightfall, even though you're not allowed to enter or clear it until morning.

Split your platoon, even if you don't think its a good idea, and here's how to split it.

Do what you're told to do.

We're too busy right now; we'll get back to you when we can.

Who specifically made each of these choices was now a moot point. That the choices had been made and the platoon was being ordered by their entire chain of command to obey them was not. As leaders, the choices we make in the context of the moment not only shape our individual paths, they also shape the paths of the people for which we have the privilege to lead. In the context of the moment as Serial #2 sat at the Y-intersection, the sum total of choices inflicted upon them by their chain of command were compounding, one on top of the other, to create a perfect storm of chaos on the horizon in front of them. Despite their best efforts to avoid it, the platoon's stressed out and unwitting brains were about to drive directly into it.

6:32 P.M., Slot Canyon: As the sun set behind the mountains, Serial #2, with its nineteen Rangers in six vehicles (four Humvees, the Jinga truck, and the inoperable Humvee), entered the mouth of the same narrow slot canyon that Serial #1 had passed through a few minutes earlier. With the Jinga truck now leading the way, Serial #2 crawled forward with extreme caution.

Ranger Machine Gunner in lead Humvee: "The canyon was the most foreboding terrain I had ever seen. I wasn't scared in the traditional way; I was more worried about how vulnerable we were. After they told us we couldn't blow up the Humvee and ordered us to tow it instead, we lost all respect for our chain of command. Once we started moving out of Magara, we were talking about what happened, and our squad leader told us to put it

Close-up view of entrance to slot canyon as seen from machine gun turret of Humvee.

behind us and focus on the mission at hand. It wasn't easy to do. Here we were following behind an Afghan Jinga truck, driven by a civilian, who is dragging a Humvee that was so badly damaged it could never be used again. As hard as we tried to put the anger and frustration with their decision to split the platoon and continue towing the Humvee out of our minds, it was impossible to do because it was right in front of us. We had to live with it, and no matter how hard we tried, we couldn't see anything good about it. Then our squad leader snapped us out of it again when he said, 'Let's stay alert; it would be so easy to take all of us out. So easy."

Boom. A minute or so after the words left the squad leader's mouth, a radiant white flash erupted on the cliff forty feet above. The convoy stopped in its tracks. A plume of smoke billowed skyward as a boulder the size of a vehicle began tumbling, end over end, downward toward the convoy. All the Rangers could do was watch and get ready to dive out of the way.

The boulder landed between the 3rd and 4th vehicles and rolled harmlessly to its new resting place in the wet sand of the creek. The

General location of first explosions and gunfire. View from lead vehicle of Serial #2 looking northeast.

platoon sergeant who was riding in the rear vehicle initially believed the explosion to be an IED, so he got on the radio and ordered everyone to "dismount the vehicles." The Jinga truck driver had already slammed on his brakes. The two Rangers who were riding with him immediately jumped out and started returning fire at what they believed were targets along the northern ridgeline. Then two more rounds exploded, and muzzle flashes blazed from the cliffs above. The slow, deliberate sound of automatic weapons fire followed. The canyon erupted as the Ranger machine gunners returned fire.

6:34 P.M., Serial # 1, Canyon Road, North of Mana: After arriving in the general vicinity of Mana, the six vehicles of Serial #1 pulled over on the side of the creek bed so the platoon leader could look at his map and try to figure out where exactly Mana was located. Squad Leader #1 would later tell investigators: "We were trying to get to Mana, and we'd taken a wrong turn."

By counting and keeping track of the tributaries you've passed, you always have a good reference point for figuring out exactly where your vehicle is on the map.

While Serial #1 was still sitting in their vehicles looking at their maps, they heard the initial explosion and gunfire from the canyon behind them.

Platoon Forward Observer: "We were in our vehicles doing a map check when we heard the initial explosion, and [Squad Leader #1] was like, 'What the hell was that?' Me and the RTO [radio operator] started trying to make comms with higher, and the platoon leader pulled out his GPS to get a grid."

Platoon Radio Operator: "The PL [platoon leader] got out, and then [Squad Leader #1] got out and took off. I didn't see him again until the firefight was over. The PL grabbed me. He asked me if I had SATCOM. I said roger that, Sir, and he took off. But what the PL didn't realize is I still had to get the SATCOM antenna because it was hooked up to the vehicle, so I had to get out my mobile one. So he took off and left me, and I was setting up the antenna. Then he realized he'd forgotten me and ran back, snagged me, and I grabbed the radio, and we moved east on foot through the village."

"Don't be in a hurry to die."

Squad Leader #1: "After hearing the explosions, everybody dismounted the vehicles, and I directed two individuals to stay with the vehicles; I don't remember their names. **I told the platoon leader that I am going to try to push past the village and see if I can overwatch movement for Serial #2 out of the ambush zone.** At that point in time, I had my squad in movement formation, and then Tillman had just one Ranger with him, and it seemed like he picked up an AMF soldier with them. They were moving to the rear of our teams. I don't think there was any coordination between the two. So we moved into the village—or up to the village, which took, I would say, anywhere from three to five minutes to get to the village, because it was up a pretty steep slope."

Squad Leader #1 moving from vehicles toward path that leads to the stone house and spur. Pathway is visible on left, stone house is in the background. Behind the stone house is the northern highground. (Photo taken during investigation #4 reenactment. 2006)

Ranger Private on Corporal Tillman's team: "When the explosion went off, I wasn't sure what it was, and someone yelled, 'I think Serial #2 is under attack.' Then I saw Pat jump out of his vehicle with his weapon and run over to grab me. **He said, 'Let's go kill the bad guys,' and pretty much, 'Let's go help our boys.'** And then he started moving, and wherever he moved, I went. While we were running up the path, I noticed one of the Afghans was following us; this was unusual because most of the Afghans we worked with weren't very motivated to do anything, especially when it came to making contact with the enemy."

After bolting from the vehicles, Squad Leader #1 (along with ten other Rangers and one Afghan) followed the path of least resistance upward toward higher ground. The path had been paved by the feet of local inhabitants over hundreds of years as they walked back and forth from their homes on the high ground to the creek bed down below.

Based on the partial radio transmission that he had heard a few minutes earlier, Squad Leader #1 understood that Serial #2 was probably somewhere directly behind them and likely under attack. Yet because the earlier transmission hadn't been shared, some of the other Rangers in Serial #1 weren't sure why they were running away from their vehicles toward the sound of gunfire or where Serial #2 was located at that time.

Platoon Forward Observer: "Getting up on top of that spur was an all-time smoker, and I kept wondering, 'Where are we going?' When we got to the top of the spur, that's the first time I knew it was Serial #2 in the canyon."

In combat zones, the first priority of all leaders is to establish a common sense of shared purpose amongst their team. When it comes to purpose, there is no such thing as need to know, only need to share.

The spot in the creek where the Rangers left their vehicles was 6,000 feet above sea level. The effects of altitude on the human body start at 2,000 feet above sea level. Each Ranger was wearing a Kevlar helmet, Kevlar body armor, their weapon, a camel backpack full of water, and carrying night vision goggles, grenades, a squad radio, and miscellaneous sustainment gear for a combined weight of over fifty pounds.

When it comes to choosing whether to wear personal protective gear such as Kevlar helmets and Kevlar body armor, every situation presents a trade-off between risks and benefits. Wearing a Kevlar helmet (three to four pounds) provides increased protection against bullets and bombs, but it noticeably decreases your ability to hear, and the weight makes it more difficult to lift your head up and look around. Wearing body armor (twenty-five pounds) also provides increased protection from bullets, but it noticeably decreases your body's ability to thermoregulate, and restricts your range of motion while standing, lying on the

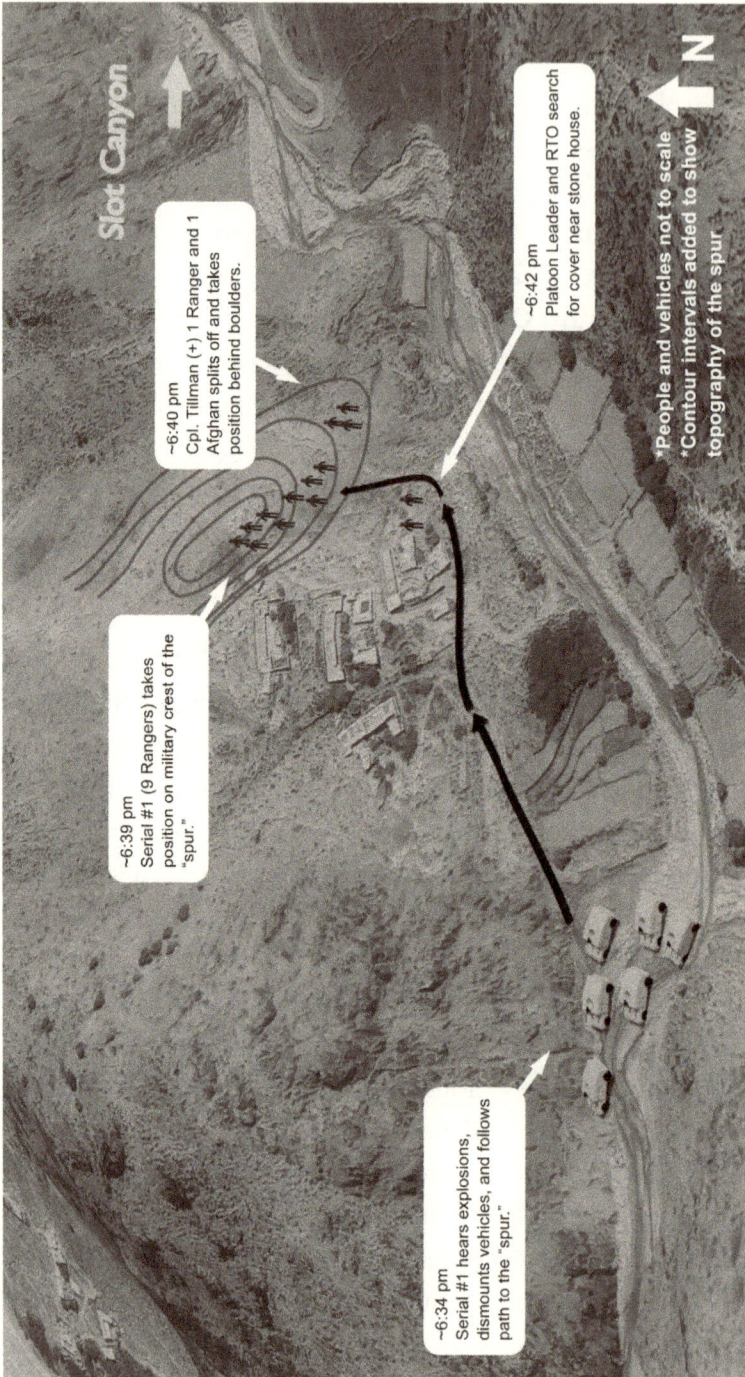

Slot Canyon

~6:40 pm
Cpl. Tillman (+) 1 Ranger and 1
Afghan splits off and takes
position behind boulders.

~6:42 pm
Platoon Leader and RTO search
for cover near stone house.

*People and vehicles not to scale
*Contour intervals added to show
topography of the spur

~6:39 pm
Serial #1 (9 Rangers) takes
position on military crest of the
"spur."

~6:34 pm
Serial #1 hears explosions,
dismounts vehicles, and follows
path to the "spur."

N

Serial #1 hears explosions in canyon, gets out of vehicles and follows path to spur ground, or maneuvering over obstacles such as walls and rocks. The Ranger regiment had a strictly enforced SOP that body armor and Kevlar helmets must "absolutely be worn whenever contact with the enemy is possible."

I chose not to wear a Kevlar helmet or body armor in Afghanistan. I was taught and learned through experience that **"speed is security" and "light is the best way to fight."** When you get caught in an ambush or any type of firefight, your survival depends on your ability to move with speed and agility. Either to a position that provides sufficient cover and concealment so you can get out of the kill zone and shoot back, or to a position that enables you to flank the enemy so you can neutralize the threat.

Much testing has been done by the Army on the effects of heavy loads on soldiers in combat. The physical limitations are obvious: it takes longer and is much harder to move when burdened by a heavy load. What's not as obvious are the psychological limitations.

When a soldier is physically fatigued and consciously aware of their limited speed and agility, they are less likely to take on added physical challenges. Thus, when a soldier is weighted down by fifty-plus pounds of gear and is already fatigued, they are far more likely to choose to stay where they are instead of moving to a safer position or one that will enable them to flank the enemy.

Note: Even though I chose not to wear a Kevlar helmet or body armor in Afghanistan, some of my fellow operators did. Sometimes it makes sense to wear them (e.g., when driving around in a Humvee with no doors, windows, or roof; or when assaulting a structure with known enemy inside). Other times, such as when you are moving through mountainous terrain, and/or fighting against small, unconventionally trained and organized enemies, it makes sense to stay as light and nimble as possible. In the Unit we had the freedom to choose what to wear and when to wear it based on our own risk-benefit trade-offs, and the reality of the situation going on around us. Unlike other special operations units, the Rangers don't have freedom to choose what to wear on combat missions and when to wear it. The lessons learned from this battle provide a timeless reminder for why they should.

After a few minutes of climbing, the Rangers of Serial #1 came to a stone house with a couple of smaller stone structures surrounding it. According to one of the Rangers bringing up the rear: "it looked like they'd been there for a thousand years." The Rangers could tell it was occupied because "there were a couple of goats munching grass off to the side." The path kept going, so they kept following it until they came to the top of a spur where the path petered out. "Everyone was out of breath, and it looked like a laser light show up ahead, so we had to stop and take a knee. Then someone yelled, 'Did you hear that explosion?' and I thought I heard it too."

Squad Leader #1: "While we were moving up the path, I could hear Serial #2 talking on my radio, but they still couldn't hear me, so I decided we should stop moving so we could make comms with them and let them know where we were. I tried to get comms with Serial #2, but every time I'd make a transmission, it would be stepped on by somebody.

"My intent was to spread the squad out along the spur [*see picture*]. The way my command came out was, 'I want the squad arrayed on the shallow edge of this spur.' My two team leaders know that means Alpha Team on the left, Bravo Team on the right; they pick out the final locations while I center myself between them. That's the way we always operate as a squad."

What Squad Leader #1 was calling "the shallow edge" of the spur was what is known as the "military crest" of the spur. Military crest refers to the area on the reverse slope of a hill or ridge that is located just below its highest point. Its importance for survival

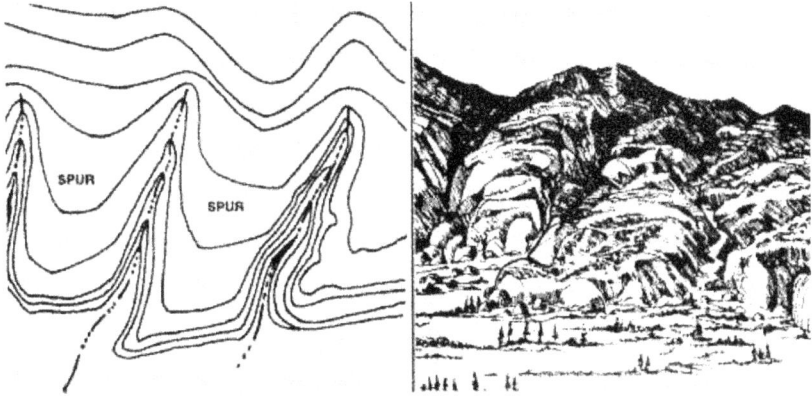

A spur is a short, continuously sloping line of higher ground,
normally jutting out from the side of a ridge like a tongue.
A spur is often formed by two, rough-parallel streams,

is a matter of simple physics. The military crest offers maximum protection and concealment from direct-fire weapons and those who are aiming them at you.

Training and fighting in mountains establishes an instinctive awareness of the military crest on any and all terrain features encountered. A soldier only has to experience being shot at or being detected once to recognize the military crest's foundational importance for survival.

Like a well-oiled machine, the Alpha and Bravo team leaders, both sergeants (E-5s) themselves, immediately began positioning their teams along the military crest in accordance with Squad

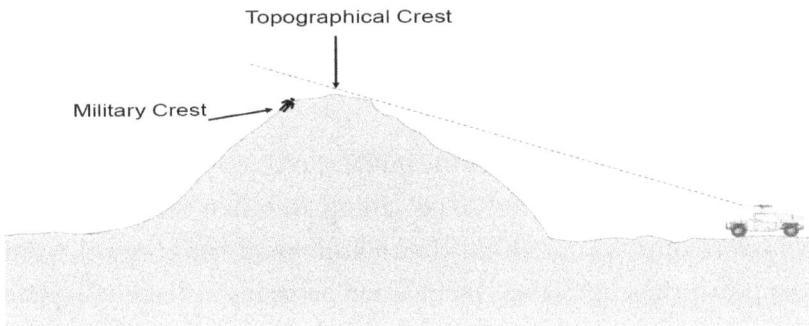

Leader #1's guidance. This left Corporal Tillman, who was newly attached to the squad and not privy to how they "always" operated, to try to figure out where he should fit in.

Squad Leader #1: "Corporal Tillman said to me, 'I spotted enemy on the high ground to the southwest, is it okay if I take my team forward toward the big rock where there's better cover and concealment?' I stood up so I could see the rock and then told him, 'Roger, go ahead.' So Corporal Tillman, along with a Ranger private and an Afghan soldier, continued moving another twenty to twenty-five meters down the southern side of the spur until they came to a couple of six-foot-long, irregularly shaped boulders, where they stopped and took a knee. All the while, I continued to try to get ahold of Serial #2 to advise them of our location and that we were going to provide fire support."

Ranger Private: "As soon as we got to the boulders, an enemy fighter began shooting at us from the opposite side of the canyon across the road to the south, so Corporal Tillman directed us to fire at the enemy position. I saw movement on the southern mountain, but the Afghan didn't speak English, so we used hand and arm signals and the barrels of our weapon to point things out. I motioned for him to come over next to me behind the boulders, but he just walked out in front of us and started firing. Pat also tried a couple of times to get him to come back behind the boulders. I don't know if it was because he didn't understand us or if he just didn't want to do what we were telling him."

Serial #2 Inside the Canyon: Riding in the lead Humvee, directly behind the Jinga truck, Squad Leader #2 described what happened next.

Squad Leader #2: "After the initial explosion on the north side of the ridgeline, I noticed rocks falling and then saw the second and third mortar rounds hit. Then a little small arms fire also from the north ridgeline. Then the platoon sergeant ordered everyone

Boulders that Corporal Tillman, the Ranger private, and the Afghan soldier use as a firing position. Note the view of the creek bed as it leaves the slot canyon (upper right) as well as the stone wall that encircles the poppy field.

to 'get back in the vehicles' so we could drive out of the kill zone as quickly as possible."

Actions upon enemy ambush is a battle drill that is trained and ingrained in every new Ranger's head. Its most important step is the first one: when caught in an ambush, your first priority is to get out of the ambush. Which is what every Ranger in Serial #2 wanted to do at that moment. The best way to describe how vulnerable the Rangers were in that canyon is to consider that the enemy didn't need bullets to kill them; they could have also used rocks.

Ranger Machine Gunner in Lead Humvee: "When the first explosion went off in the canyon, everyone stopped. Then there were two more explosions on the north side of the canyon and the guys in the Jinga truck jumped out and ran into the rocks for cover. That's when I realized 'we're trapped,' and I felt panic set in. We were in the wrong place at the wrong time. I saw muzzle flashes and movement on the high ground to the north, but I wasn't sure if I should continue covering my sector or swing around and engage the northern ridge. My teammate on the fifty-cal started firing, so I followed his lead and started firing too. [Squad Leader #2] wasn't our regular squad leader. They attached us to him at the start of the mission. All of us knew him, and he was really easy to work with, but we never conducted vehicle live-fire training together, so he probably didn't know what machine gun teams were supposed to do in a moving firefight or what commands he was supposed to use to control us."

Note: Squad Leader #2 was a rifle squad leader, not a weapons squad leader (see org chart), which normally is no big deal (the machine gun teams are usually attached to the rifle squads for dismounted missions). However, the 2nd Platoon was in a situation that no one in the platoon had ever been through before (a twelve-minute-long, rolling, vehicular ambush while trapped inside a slot canyon). Critically, they were about to transition into a situation with both friendly and enemy targets that required precision command and control of the machine gun teams. Such precise control requires knowledge of how the teams operate (tactics and techniques) and, most importantly, knowledge of the individual team members' strengths and weaknesses in executing those tactics and techniques.

Once the NCOs got everyone back in their vehicles and the drivers stepped on the gas, they instantly recognized they had a problem. The Jinga truck driver, who was leading the convoy, wasn't trained on what to do in an ambush. More than one Ranger in the convoy suspected the Jinga truck driver's refusal to move was part of a premeditated enemy plan and that he was purposefully blocking their path. The two Rangers riding with him knew better. The driver was frozen in fear and not sure what they wanted him to do because they were screaming in his face and the interpreter wasn't able to keep pace. Eventually, he got the message and hopped back in his truck to start driving. Slowly but surely, the Jinga truck crept forward.

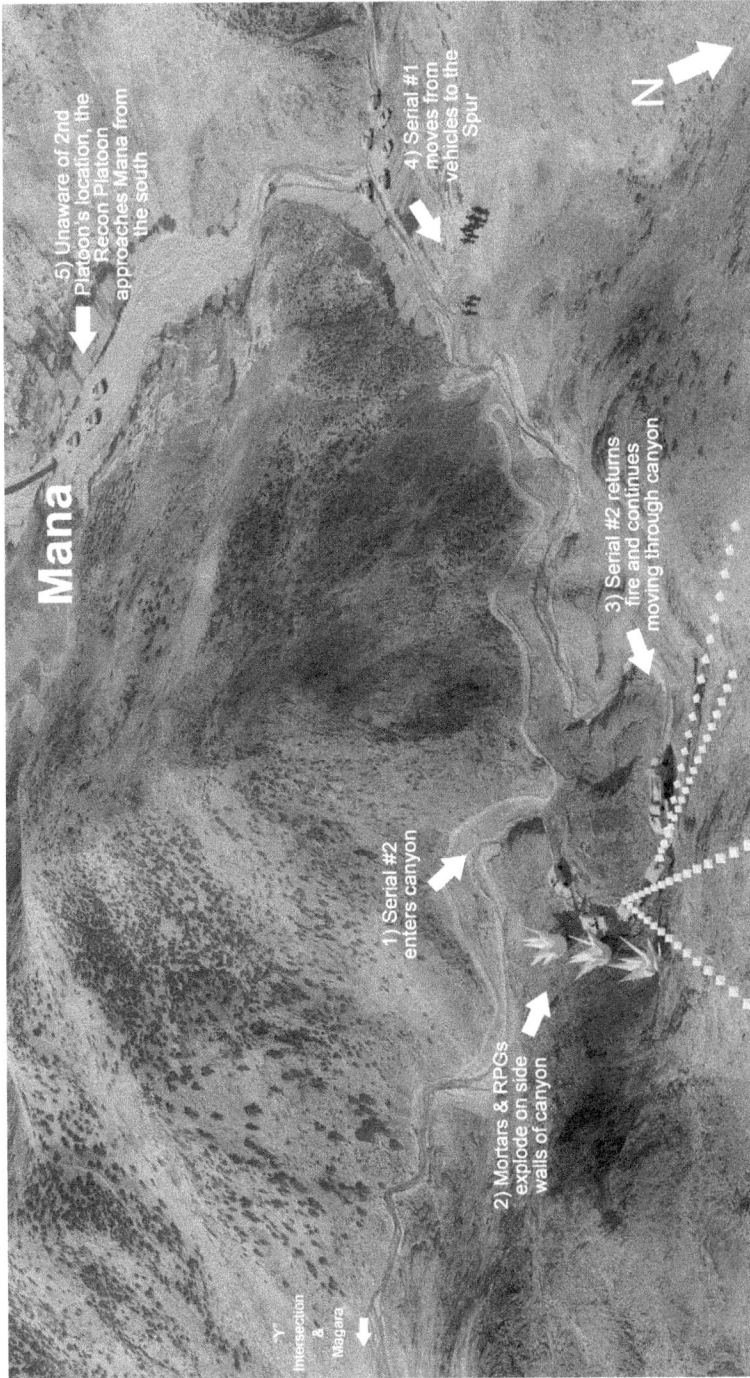

View from enemy position on northern high ground.

Getting a five-ton Jinga truck that's towing an inoperable Humvee through a narrow-necked canyon would be difficult under any circumstances; getting it through quickly while under enemy fire with heavily armed soldiers screaming at you would be almost impossible.

The river that had cut this canyon was currently a stream flowing downhill in the same direction as the platoon was traveling. The water hadn't cut the canyon as a vehicle path. It had cut a path of least resistance between giant geometric boulders and twisted rock outcroppings with walls that were one hundred and fifty to fifteen hundred feet high.

The initial explosion occurred just as the Jinga truck was approaching one of the narrowest points of passage. (See picture.) The Jinga truck driver had to thread the needle to get both his truck and the inoperable Humvee all the way through it. He proved he was up to the task, but the boulder-strewn creek, undulating terrain, and constant hairpin turns meant there was no way to drive at a speed much faster than a man can walk. And it wasn't just the Jinga truck that was having difficulty. As one of the Rangers who was riding in the rear vehicle with Kevin Tillman described to

One of many hairpin turns inside the canyon.

investigators, "the driver took a turn too sharp, and it [the canyon wall] ripped the buttstock off my machine gun."

Lead Humvee Driver: "Getting the Jinga truck driver to move was extremely difficult. It was stop-and-go after the first explosion for most of the way. And then, halfway through the canyon, at another one of the narrowest points of passage, the Jinga truck driver came to a complete stop. So we dismounted a second time, and it was at that time that our squad leader positively identified another enemy pax [person] on the ridge and told us to take out the AT-4 rocket and fire it at him. Everyone in the vehicle was firing on known, likely, and suspected locations."

The three Rangers in the back of the lead Humvee continued to engage enemy targets on the northern ridge with .50 cal., M240B, and M249 machine guns. Concurrently, several of the Rangers in the vehicles behind them spotted "what appeared to be three enemy combatants running along the northern ridgeline, which was located approximately one thousand meters from the canyon." The Rangers quickly dismounted and began firing everything they had at the enemy targets as well. One of them told investigators he "fired his entire basic load of forty-millimeter rounds from his M203."[74] The mortar section also dismounted, quickly set up, and fired "one round at CH 6,[75] two hundred meters, on the north ridge, to get a good range on a target should they see one." The snipers also engaged targets on the northern ridge using their SR-25 sniper rifles and long-range Leupold sights.

6:39 P.M., West of Spur, Serial #1: The platoon leader and his radio operator continued sprinting up the same path that Squad Leader #1 and his crew had followed a few minutes earlier. As they approached the stone house where the others had seen some goats, the radio operator spotted an enemy position on the northern ridge, so he and the platoon leader stopped running and fired "a half magazine before taking a knee to catch our breath."

View of the stone house. Inside the white circle are two role players and a rucksack with a radio in the same location as the platoon leader and radio operator when they were knocked unconscious by indirect fire (e.g., mortar or grenade). The spur is located 70–80 meters to the right/east of the stone house.

The stone house seemed like a safe place to set up their radio so they could send a sit-rep (situational report) back to the chain of command in Khowst and start communicating with the rest of the platoon. **They were able to quickly call in a "troops in contact" report to HQ,** and the radio operator described what happened next.

Radio Operator: "The rate of fire all around us started to pick up, and I said, 'Hey, Sir, we can't stay here, we got to move,' because, I mean, rounds were hitting the side of the building, I could see rounds kicking up all around, and I was like, 'Hey, Sir, we got to move.' So, he was like, 'Roger that, where do you think we should move to?' and I kind of briefly looked around, and I was like, 'I don't know, Sir, I have no idea.' And that's when—there was a large explosion—what, at that time, I thought was a mortar round…the mortar round exploded, blew me basically out away from the wall onto my face. I had shrapnel, small pieces in my arm and in my neck."

Both the platoon leader and the radio operator were "knocked senseless" by the blast overpressure.[76] As a result, the platoon leader was incapacitated and no longer able to effectively command and control the platoon. Their position next to the stone house was almost exactly two hundred meters from where the forty-millimeter grenades and sixty-millimeter mortar had been fired a few seconds earlier by Serial #2.

6:40 P.M., Outskirts of Mana, Reconnaissance Platoon Sergeant: "We were driving on the dry creek road, one terrain feature over from the 2nd Platoon—though we had no idea they were there. It's always a bumpy ride in the creeks, but on this day it was super bumpy. Three of our rucksacks came loose and fell off the back of one of our vehicles just as we were approaching the town of Mana.

"When we stopped to put them back on, we noticed that all the women and children in the village were standing on their rooftops. This was odd because it was sundown, and sundown is when Afghans are in their homes either praying or eating. There were no fighting-age men in sight. A few minutes later, while we were reattaching the rucksacks to our vehicle, four or five of the creepiest-looking Afghans any of us had ever seen walked up to us. All were young men, and they were dressed in really weird clothes, like they were from somewhere else. Our interpreter started talking to them, and he got real nervous because they were pretty hostile. They kept saying to him, 'Why are you here? It is very dangerous, you shouldn't be here, you need to leave.' Once we had the rucksacks secure, we got back in our vehicles, and **that's when we heard the 'troops in contact' call from the 2nd Platoon, we could hear gunfire in the background as he talked.** We checked their location on FalconView[77] and were shocked that they were only a few minutes north of us.

"We approached their location with extreme caution. One of the most difficult and dangerous things to do in combat is to link up with a unit in contact. To this day, I believe that if those rucks hadn't fallen off and made us stop, we would have driven directly into Serial #1's location, and since neither of us knew the other was there, who knows what would have happened."

According to the officer who wrote the first draft of the first investigation: *"At approximately 6:40 P.M., the TOC received the 'troops in contact' report from the platoon leader's RTO [radio operator].* **There is also evidence that the battle staff was surprised by the report because there was initial confusion about 2nd Platoon's movement during daylight hours.*"*

What follows are sentences inserted by someone in the 2nd Ranger Battalion Headquarters who reviewed the draft:

"In the paragraph above, you said 'the TOC' was 'surprised' by the report, but 'WHO' is the TOC? So, I took some liberty here with the wording: The TOC responded by immediately processing requests to support 2/A with aerial observation, close air support with AH-1 attack helicopters from Bagram, A-10 attack aircraft, and a UH-60 MEDEVAC helicopter."

6:40 P.M., Slot Canyon, Serial #2: When you're caught in the middle of an ambush kill zone you have two options: to flee or to fight. With the Jinga truck and steep terrain taking away the option to flee, the only other option was to fight. The Rangers of Serial #2 were trapped.

Most of the Rangers in Serial #2 told investigators that the Jinga truck stopped three times, and each time it stopped, they had to dismount their vehicles. When the vehicles weren't stopped, they crawled. It was like attacking in slow motion.

Lead Humvee Machine Gunner: "When we stopped the second time, I was kind of amazed we were still alive, and I guess you could say with each passing second I went from being convinced we were going to die to maybe these guys don't know what they're doing to maybe all our suppressive fires are working. So we kept talking the guns."

Note: "Talking the guns" refers to a technique where machine gun teams work together by alternating their fires at enemy targets. In this case, the first machine gunner would fire a 10-round burst from his M240, and then his teammate would respond with a 10-round burst from his .50 cal. **Talking the guns is one of the best ways to provide continuous suppressive fire on a target while also preserving ammunition.**

While the vehicles crawled, the drivers stayed focused on driving, and everyone else stayed focused on firing their weapons at the enemy. The Rangers in Serial #2 were firing everything they had in a thundering fusillade of thousands of rounds, which they aimed up and over the rim of the horseshoe-shaped canyon that surrounded them. Most crew-served weapon systems have three rates of fire that all soldiers are required to learn when they're issued the weapon. As an example, the M249 SAW has a sustained rate of fire of one hundred rounds per minute, a rapid rate of fire

Approximate location where Serial #2 stopped the second time inside slot canyon.

of two hundred rounds per minute, and a cyclic rate of fire of 650 to 850 rounds per minute. (See Weapons Annex for information on weapons used by 2nd Platoon.)

The rate of fire that an individual machine gunner uses is dependent on the enemy situation as well as how much ammunition is available. Thus, an M249 SAW machine gunner who is moving on foot, such as Corporal Tillman, normally carries between 600 and 1,000 rounds, which are packaged in 200-round drums. No matter how many rounds the enemy fires, the gunner is trained to constantly monitor and maintain awareness of how many rounds he has left. The more he fires, the more he adjusts his rate of fire downward to ensure he doesn't run out of ammo.

In contrast, an M249 machine gunner who is firing from a vehicle, such as those in Serial #2, isn't limited to the amount of ammo he can carry on his back or belt. Sitting within arm's reach, and all around him, are thousands of additional rounds stored in ammo cans and boxes. Therefore, the machine gunners in Serial #2 could maintain a much higher rate of fire—likely 200–300+ rounds per minute—with little worry of running out. The only pauses in their rates of fire occurred when the gunners were forced to swap out their red-hot barrels or break into brand new boxes of ammunition before reloading and firing some more.

Three separate Rangers in Serial #2 riding in three separate vehicles told investigators they "spotted enemy personnel on the tree and rock-lined ridges that rimmed the canyon." Most of the Rangers in Serial #1 told investigators they also spotted enemy personnel on the northern ridge. Although the leaders of both Serial #1 and Serial #2 were continuously trying to call each other and report what they were seeing, the close-quarter canyon walls prevented their line-of-sight radio signals from snaking their way through and allowing them to communicate.

As a result, not only did the Rangers in Serial #2 not know

where Serial #1 was in relation to their position inside the canyon, they also had no idea that Serial #1 was shooting at the same enemy targets on the northern ridge, where, as one of them described to investigators, "our rounds were all hitting."

Top of Spur, Ranger Forward Observer: "At this point, the rounds that were flying over our heads just increased dramatically. There were just tons of rounds, and I saw an explosion. And somebody screamed, 'They're behind us, they're moving on us,' everybody fired some rounds up on the hill."

Another Ranger who was positioned a few feet away from the forward observer described it this way: "As I dropped to the ground, I heard whistling and snapping and recognized that as incoming rounds. At that point, we all believed that the rounds were coming from a ledge about two hundred to three hundred meters northeast of us and about one hundred and fifty feet above us. I called out the location and heard Pat Tillman echo it; we put sustained fire on the ledge for about thirty seconds."

Now it was the Serial #1 Rangers on top of the spur who couldn't make sense of the "confusing and confounding crossfire" that was pinning them down. One of the investigators commented "it was highly likely that at least some of that fire was coming from the Rangers in Serial #2." (See map #.) Yet how could that be if both serials were firing at enemy targets in the same general area on the northern ridgeline?

To understand what was happening to the trajectory of their bullets, we have to see it through the eyes of the Rangers in Serial #2. By watching a reenactment video[78] taken during the fourth investigation, which was filmed from the perspective of a machine gunner while his Humvee drove through the canyon on the same day and time of year (22 April 2006), we can better understand what was actually happening. (See video reenactment of entire drive through slot canyon at firefight.commonsenseway.com)

The video should be used as a training film for teaching future machine gunners the perils of trying to engage long-distance targets while driving off-road with a non-gyro-stabilized weapon system. The reason it's so perilous—and not recommended—is because the movement of the vehicle (up, down, north, south, east, and west) makes it extremely difficult for the machine gunner to maintain his sense of direction and to stabilize the barrel from which the bullets are firing.

The term "sense of direction" means the direction to a known location that our senses have a reference for (e.g., the sun setting in the west, the North Star, a mountain in the distance, etc.). To calibrate and update our sense of direction, grid cells in the neocortex fire in a repeatable pattern based on where we are in relation to the known location. However, without such a reference point or when the head is unexpectedly turned, facing multiple directions in a short period of time, our brains can't calibrate and we lose our sense of direction.

When the vehicles originally entered the canyon, they were heading west; then the canyon turned northwest, then north, then northeast, then north again, then northwest, then west, then southwest, then south, then southeast, then back south again, then southwest, and finally, as they exited the canyon, directly west.

To navigate the horseshoe canyon and serpentine creek bed, the drivers not only had to make macro-directional changes to compensate for the cardinal direction changes of the canyon's path, they also had to make micro-directional changes—like a race car driver—zigging and zagging their vehicles around the creek bed to avoid the massive rock outcroppings that defined its path.

The combination of a bumpy ride and the inability to maintain their sense of direction resulted in wild swings in both the horizontal and vertical directions of the machine gun barrels, which increased the diameter of the "cone of fire" as well as the circumference of "beaten zones" where the bullets landed.

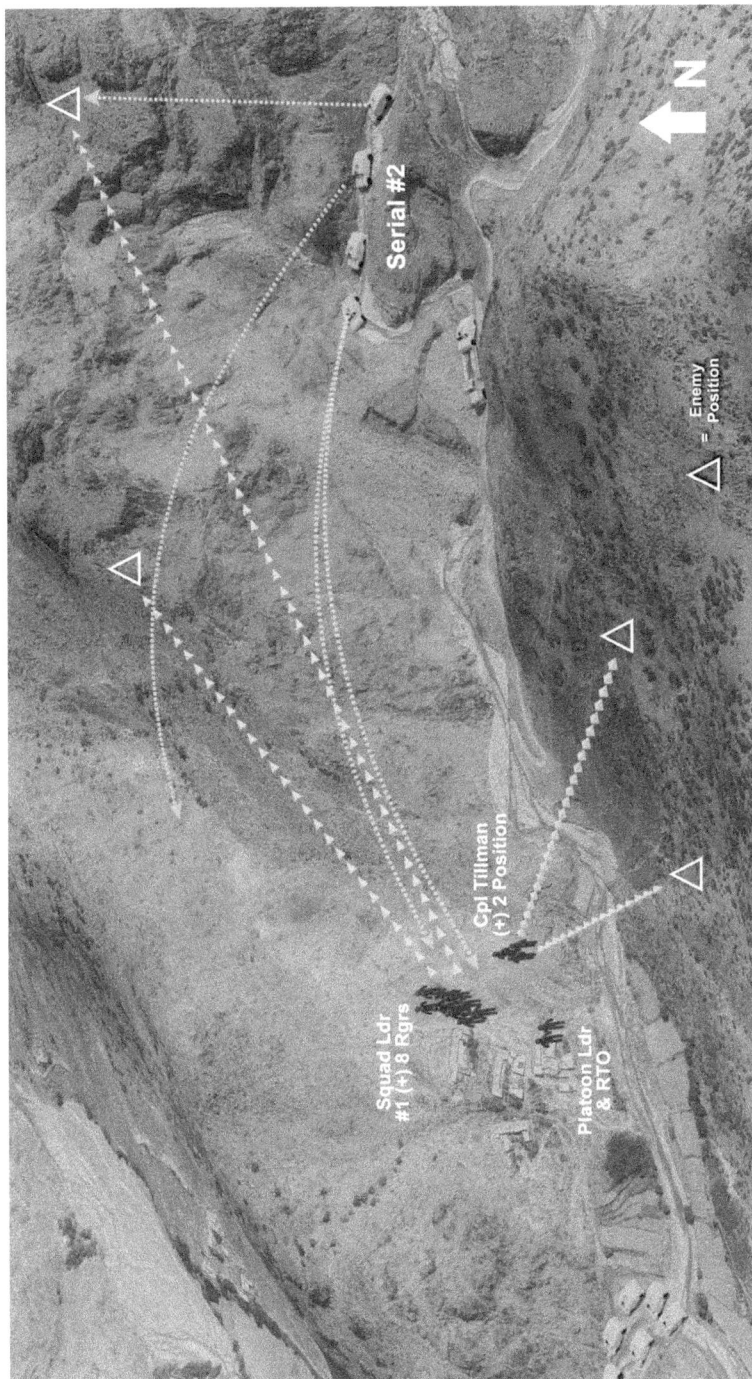

Crossfire in the canyon as the two serials were about to converge.

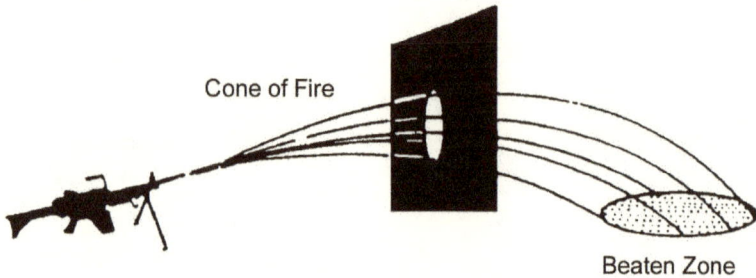

Cone of Fire

Beaten Zone

The "cone of fire" is the pattern created when several rounds are fired in a burst from any machine gun, as each round takes a slightly different trajectory on its way to the target. This pattern is caused primarily by the vibration of the machine gun barrel and the platform or vehicle it's mounted on, as well as variations in ammunition and atmospheric conditions.

The "beaten zone" is the elliptical pattern formed by the rounds within the cone of fire as they strike the ground around the target. As the vibration or movement of the barrel increases, so do the size and shape of the beaten zone. When firing on the move at long-distance targets, the beaten zone from high-caliber weapons such as the .50 cal, and M240B can be over a hundred meters long.

On the Spur, Serial #1: After engaging targets on both the northern and southern ridgelines, Corporal Tillman told the Ranger private who was crouched beside him, "Stay here. I'm going to link up with the squad leader to get further guidance." After dashing up the spur, Corporal Tillman took a knee alongside Squad Leader #1.

Squad Leader #1: "He asked me for permission, which I thought was kind of out of the ordinary, to drop his body armor so he could maneuver faster to a position on the southern ridgeline [across Canyon Road] where he had spotted enemy positions. I told him not to drop his body armor and that Serial #2 didn't know where they were, so he should stay where he was at and continue to provide suppressive fires."

While Corporal Tillman was talking to Squad Leader #1, the platoon forward observer was lying on the ground a few feet away. When I asked him about the conversation, he told me: "The last thing I heard Corporal Tillman say was, '**I can't leave my guys out there**,' before he dashed back to his position behind the boulder on the eastern side of the spur."

The Ranger private next to Corporal Tillman was asked by investigators what Corporal Tillman told him when he got back from talking to Squad Leader #1: "I continued firing until he came back, and when he came back, he told us, he told myself what we were going to do, and at that time, we started taking fire again from the south from that point on. Yes, Sir."

Investigator: "Where was he firing at? Could you see the enemy locations?"

Ranger Private: "He was shooting right up on top here [across Canyon Road on the high ground to the south]. I could see the top of the ridgeline, and I saw muzzle flashes, and I just saw where he was shooting, so that's what I assumed."

Slot Canyon, Serial #2: When they first drove into the canyon, many of the Rangers described it to investigators as a funnel. Once they started driving through it, some described it as a maze. After blasting away continuously for almost ten minutes, at least one of the Rangers described it as "a megaphone."

Hearing plays a vital role in the success or failure of soldiers and units in combat due to its critical importance to speech processing. Noise-induced hearing loss is a significant impairment that can seriously degrade individual and team performance in combat. Military personnel are constantly exposed to high levels of noise throughout their careers (gunfire, explosions, aircraft engines, etc.), so it's not surprising that noise-induced hearing loss and tinnitus (constant ringing in the ears) remain the second-most prevalent service-connected disabilities. Much of the noise experienced by military personnel exceeds the level of maximum protection achievable with double hearing protection.

According to the National Institute for Occupational Safety and Health (NIOSH), "exposure to noise sources exceeding 140 dB for longer than twenty-eight seconds causes the delicate inner ear tissues (stereocilia) to stretch beyond their elastic limits, resulting in direct damage to the supporting sensory cells."[79] This type of acoustic trauma usually results in temporary or reversible hearing loss. The amount of time above and beyond twenty-eight seconds that an individual is exposed to a noise source exceeding 140 dB directly influences the amount of time it will take to recover (anywhere from minutes to hours to days).

To fully understand the acoustic trauma suffered by the Rangers in Serial #2 we must first consider the decibel levels of the weapons they were firing: AT-4 Rocket = 190 dB, 60mm Mortar = 185 dB, M249 MG = 160 dB, .50 caliber MG = 159 dB, CAR-15/M-4 = 157 dB. Now consider that they had been exposed to these decibel levels—most without hearing protection—for somewhere between eight and twelve minutes while trapped inside an echoing canyon.

Decibel Levels	Source Examples	Effects After 28 Seconds
0 dB	Silence	Threshold of Hearing
10 dB	Whisper	Barely Audible
40–50 dB	Bird Chirping	Quiet
95–105 dB	Jet Flying Overhead at 1,000 ft.	Moderately Loud
105–115 dB	Live Rock Band	Very Loud
115–139 dB	Military Jet Preparing for Takeoff (50 ft.)	Uncomfortably Loud
140+ dB	Military Weapons	Temporary Hearing Loss

Most Rangers in Serial #2 reported temporary hearing loss in their statements to the investigators: "I had lost my hearing, I was right by the canyon wall when the shooting started, so all I could hear was ringing;" "My squad leader had to yell in my ear because I couldn't even hear my own radio;" "While we were in the canyon, I tried calling [Squad Leader #2] and I couldn't get him on the radio at all. I think at that point he was deaf because of his fifty-caliber gunner. Because when everything was said and done, everyone who was around that vehicle couldn't hear anything. I was having to yell in their ears."

Not everyone in Serial #2 suffered the same sensory fate. As two members of the mortar section who were riding with Serial #2 would later tell investigators, "I think we were the only ones in the whole convoy who had Peltor headsets, so, I mean, we heard every radio transmission that came across, and normally the only person that was really talking on it was the platoon sergeant to us, and he had no situational awareness so he was trying to get it. But as far as Serial #1, the first time I heard them talk on the radio was after we had all stopped and [Squad Leader #1] was calling for a Medevac."

Peltor headsets (see picture) use frequency leveling technology to minimize the potential damage to hearing that occurs as a result of sudden high decibel noises such as gunfire. The protection goes beyond just gunfire. To survive and stay safe in combat, you have to continually understand what's going on around you, which includes things like the sound of enemy footsteps trying to sneak up behind you or a verbal warning from your teammate telling you that you're about to walk into their line of fire. With Peltor's frequency leveling technology, the wearer can hear verbal commands, as well as someone talking on the radio, even while shooting their weapon. When not shooting, the wearer can hear someone walking or whispering from over fifty feet away.

The majority of Rangers in Serial #2 were either not issued or not trained to wear their Peltors. As a result, the majority had not only lost one of their most valuable senses but also a key component of their neocortex's (thinking brain's) capacity to make sense of what was going on around them.

Here's how the regimental sergeant major described the canyon to investigators: "Having been in a while and been in numerous contacts, that was probably one of the most confusing I've had. I think it was the most confusing only because it occurred in that canyon. We had no visibility. It echoes really bad. You could not really tell any kind of direction of incoming or outgoing fire when you were listening for gunfire. So, when myself and others returned fire when we were still deep in the canyon, I would kind of direct mine towards the rim of the canyon to keep people back."

6:42 P.M., Slot Canyon, Serial #2: With the Jinga truck and the disabled Humvee still blocking the rest of the convoy's path, a bottleneck of vehicles began to build. Squad Leader #2, who was

Peltor Headset

riding in the lead Humvee directly behind the Jinga, "jumped out of his vehicle to confront the driver and see if he could make him move." There was "a lot of shouting," which some Rangers would later describe to investigators as "a heated fight" between the two.

Squad Leader #2 was not known as a hothead. Instead, he was highly regarded by his superiors and subordinates alike. Many of whom described him to investigators as "smart," "easy to work with," and one of "the best Ranger NCOs in the battalion." He was Kevin Tillman's former team leader, and Kevin had referred to him more than once as "totally squared away" and "the kind of leader I want to work for again someday."

According to the driver of the lead Humvee, "the Jinga truck driver refused to move, so [Squad Leader #2] had to move up the hill and break the side window and point his weapon at the driver's face to make him understand." Squad Leader #2 then jumped into the passenger seat, and a few moments later the convoy was moving forward again. As the distance between the walls began to widen, it finally appeared as if there was going to be enough room for the trapped Humvees to pass the Jinga truck.

While they inched forward, Squad Leader #2 told investigators he "spotted another enemy moving along the northern ridge, so I swung my M4 around and shattered the passenger side window of the truck to get a shot off." Once again, the driver screamed and yelled in protest as Squad Leader #2 fired his final thirty-round magazine while halfway hanging out the door. He then realized he was out of ammo and jumped out of the Jinga truck to head back to his vehicle and reload.

Perhaps it was the opening of the narrow-necked canyon walls. Perhaps it was the fact that they were now heading west. (See map.) Whatever the reason, Serial #2's ability to communicate on the radio suddenly opened up. According to a Ranger riding in the third vehicle behind the Jinga truck, "**[Squad Leader #2]** was all amped up, so the platoon sergeant, who was extremely calm, got on the radio and was telling him to 'calm down and continue moving.'"

When we speak calmly, we calm the way we act. The platoon sergeant provided a textbook example of what all leaders should do when one of their people is cortically inhibited. Unfortunately, **unbeknownst to the platoon sergeant, Squad Leader #2 had lost his hearing, so he couldn't "make sense" of what anyone was saying to him over the radio.**

6:44 P.M., Top of Spur, Squad Leader #1: "I heard over the radio that Serial #2 was mounting up to get around the towed truck vehicle. I remember seeing the lead vehicle starting to make its way out of the canyon, and I had to stand up to look over the spur. I told everybody on the fire teams that 'friendlies are coming out of the low ground and the lead vehicle was coming out of the canyon,' and they mimicked the call while I ducked back down. **When I saw the vehicle coming out, I also saw Corporal Tillman's position. I knew Serial #2 didn't know where they were.**"

Recall the relationship between experience in combat and experience with friendly fire. Once learned, our brains see the world through the pattern-revealing lens of that relationship: *They don't know where we are; They've been shooting at the high ground for the last ten minutes and when they come out of the canyon we're going to be between them and the high ground; We're not down in our vehicles where they expect us to be, we're up here on the high ground where they expect the enemy to be; Some of those guys may be amped up and cortically inhibited.*" When we combine what's going on around us with what we know, the logic of why we do what we do and choose what we choose emerges: Get down and stay down.

Of course, there was no way for Squad Leader #1 or anyone else to predict what was about to unfold. However, it is likely that Corporal Tillman and the Ranger private's organic squad leader—who at that moment was riding in the front seat of the Jinga truck, and who was intimately familiar with his men's tactical experience and knowledge—would have recognized that this was a situation that neither Corporal Tillman's nor the Ranger private could possibly comprehend.

Saying it out loud is not meant to second-guess Squad Leader #1. He didn't order the platoon to split its teams apart and arbitrarily attach Corporal Tillman and the Ranger private to his squad. **Instead, saying it out loud enables our brains to see a real-world example of what can happen in combat when a disconnected chain of command breaks apart organic team relationships when there is no operational reason to do so.**

Top of Spur, Serial #1 Ranger: "I thought an enemy mortar might have landed near Tillman's position. And then I heard a different weapon system firing that wasn't familiar to me, and it was coming from the same direction as Corporal Tillman's position."

The weapon was an AK-47, and it belonged to the only Afghan who accompanied the Rangers of Serial #1 when they bolted from their vehicles and followed the path to the top of the spur. The Afghan wasn't attached to Corporal Tillman's fire team until he did so himself at the top of the spur. In the middle of the ongoing melee, the Afghan soldier continued standing in the same spot (approximately ten feet directly in front of the boulders) while firing his AK-47 at or over the canyon road/creek bed below.

Squad Leader #1 told investigators, "The Afghan soldier was just kind of doing his own thing; obviously, I couldn't provide him with any direction." The Ranger private at Corporal Tillman's side told investigators that "both he and the AMF soldier were shooting at a possible enemy position to the south, shooting across the road." The first investigation proposed the Afghan had "either spotted the same enemy that Corporal Tillman was shooting at on the southern ridge across the road or he was shooting in the direction of Serial #2, who he mistook as the enemy."

Engaging enemy targets on the southern ridge across the road, or mistaking Serial #2 as the enemy might make sense of why the Afghan was firing his AK-47. What's harder to make sense of is why he was doing it while standing in the open, in front of the Rangers' position, and completely exposed. The fact that he didn't speak English certainly has to be considered as a major contributing factor. Additionally, he had just joined the platoon for the first time that morning, so the men did not have firsthand knowledge regarding his past training and experience, nor insight into his mindset at that specific moment in time.

Why is that important? Because while not speaking English explains why he didn't understand what the Rangers were saying to him, it doesn't explain why he failed to comprehend the message being delivered via the international language of bullets whizzing over his head and projectiles exploding all around him. Everyone else got the message loud and clear: "Get down behind cover." The Afghan continued standing and shooting.

As the Ranger private explained to me: "Pat and I were both getting frustrated because we realized the Afghan soldier was probably drawing enemy fire toward us." **What they had no way to comprehend was that he was about to have the same effect on friendly fire.**

From a combat leadership perspective, it is valuable to put ourselves in Corporal Tillman's shoes at this moment in time.

When Serial #1 heard the mortar round explode in the canyon behind them, Corporal Tillman grabbed his M249 machine gun and told his teammate: **"Let's go kill the bad guys,"** and **"Let's go help our boys."** After coordinating with Squad Leader #1 on top of the spur, the last words he said before running back to his team were, **"I can't leave my guys**

out there." Once he was back behind the boulders, he was on his own and isolated from the rest of Serial #1 and from the other members of his organic squad.

As a team leader, he was responsible for the two men who were with him at that moment. One of them didn't speak English and didn't seem to understand basic tactics. The other was the newest and youngest member of the platoon. His performance over the previous eight days had earned him the trust and respect of Corporal Tillman and his fellow Rangers. However, like all "new guys," the Ranger private's lack of experience made him dependent on his team leader—Corporal Tillman—for direction and guidance on what he was supposed to do next.

Bravery and tactical acumen have a lot to do with the company we keep. When you are surrounded by highly trained and experienced teammates, you are able to fully immerse yourself in what's going on around you and bounce what you are seeing and thinking off of them to see if it makes sense. When Corporal Tillman suggested dropping his body armor and crossing the road to take up a position on southern high ground, the idea may have sounded "out of the ordinary" at first thought. Until you realize that his recommendation was borne of the fact that their current position on the spur wasn't doing any of the things that Squad Leader #1 and the platoon leader assumed it would do when they first ran up there.

Serial #1's position on top of the spur didn't enable them to talk to Serial #2, so they couldn't tell Serial #2 where they were. And the terrain around the spur didn't allow them to provide fire support for Serial #2 while they were inside the canyon. **From a risk-benefit perspective, why stay in a dangerous place—far away from their vehicles—that provided no tactical benefits and came at a huge cost in the way of exposure to both enemy and friendly fire?**

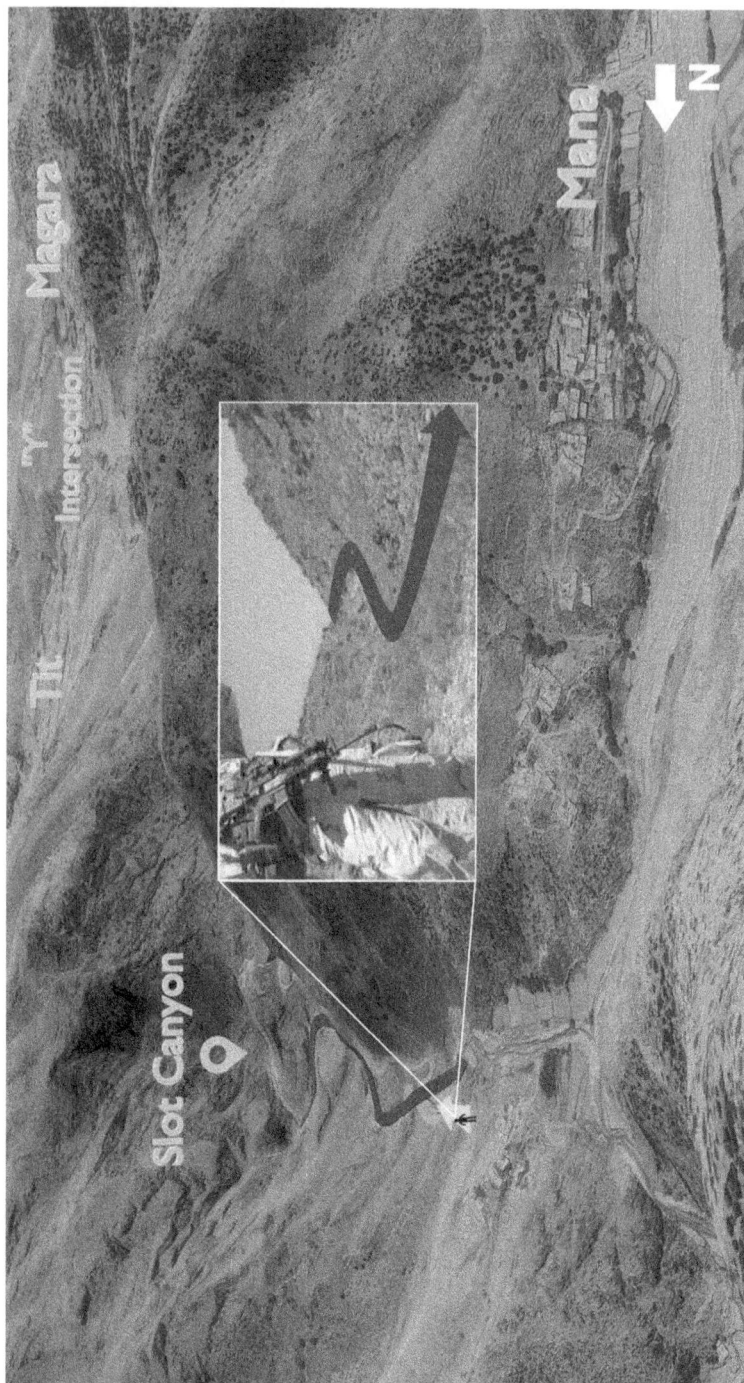

Serial #1's position on top of the spur (looking east) didn't enable them to see Serial #2 in the slot canyon (the black line marks the ground path of Canyon Road). Additionally, the towering terrain prevented their radios from establishing line-of-sight communications.

Sometimes moving only twenty meters enables a radio to establish positive communications. Based on line-of-sight terrain analysis of the area around the spur, moving twenty to fifty meters south of the boulders to the vicinity of the rock wall would have enabled them to establish communications with Serial #2 while they were still trapped inside the canyon. And as we now know in hindsight, anywhere else would have been safer than their current position away from their vehicles, up on top of the spur.

If Corporal Tillman's team had been down near the road (where Serial #2 expected them to be) when the lead Humvee emerged from the canyon, it is likely he would have had a much better chance of signaling them to cease fire and alerting them that there were friendlies up on the high ground.

I admired Corporal Tillman's suggestion to drop his body armor and move his team south in order to improve Serial #1's options the very first time I read about it. **Instead of panicking and/or doing nothing, Corporal Tillman was thinking, and in the process, he displayed one of the hallmarks of a great combat leader: common-sense initiative.**

Statements made by the other Rangers on the spur reveal that as the bullets were flying over his head from multiple directions and the Ranger private next to him was hanging on his every word, Corporal Tillman remained cool, calm, and collected while firing hundreds of rounds at multiple enemy positions. As you'll read in the pages that follow, he was singularly responsible for keeping the young Ranger private next to him alive and uninjured.

Note: It is my belief that in the context of this moment, Corporal Tillman's combat leadership was exemplary and more than sufficient to earn him recognition with a combat valor award such as a Silver Star.

Western Side of Spur: The platoon leader and his radio operator were still shaking off the effects of what investigators concluded was either a sixty-millimeter mortar or forty-millimeter grenade blast. As they partially regained their senses, they realized they needed to make radio contact with the rest of the platoon and figure out what was going on around them.

Platoon Radio Operator: "When I switched to the platoon frequency, the only thing I heard was a lot of people screaming and stepping all over each other. Then the screaming suddenly stopped, and I heard the platoon sergeant say, 'Is anyone hit?' and someone came back on the radio and said, 'Fuck no,' and then they said, 'All right, let's get the fuck out of here.' And it was right after I heard him say that on the radio that the gun truck came roaring out of the canyon, and that's the first time I realized they were actually moving down this road."

Although he was riding in the same Humvee as the platoon leader, the radio operator did not hear the partial transmission from the platoon sergeant. "The platoon leader had the handset up to his ear so I didn't hear the call from the platoon sergeant. I had no idea that Serial #2 was moving behind us on Canyon Road."

Slot Canyon, Serial #2: If you've never been stuck in the kill zone of an enemy ambush, it's hard to describe the feeling. Instead of trying to describe it myself, I asked one of the Rangers in Serial #2 to tell me in his own words how he felt: "All I could think of was 'we have to get out of here or we're all going to die.' But we couldn't go anywhere because the Jinga truck and the broken Humvee were blocking us. The enemy had the high ground, so we felt like fish trapped in a barrel. It was frustrating, and there was nothing we could do about it. I was red-hot pissed off, like road rage."

After hearing the references to being "trapped" and "road rage," I better understood the mindset of Squad Leader #2 when he had his final confrontation with the Jinga truck driver. As mentioned, Squad Leader #2 was considered smart, squared away, and one of the best NCOs in the battalion. In the context of the moment, as Serial #2 was finally breaking free from the confines of the closed-in canyon, he wasn't himself. He was amped up and cortically inhibited. Worse, the contagion of his amped-up emotions was spreading like wildfire to the Rangers around him.

In life-or-death survival situations such as combat and natural disasters, the contagion of fear and panic spreads like wildfire, as do courage and calm.

It is a primary responsibility of all leaders (the highest ranking people aren't the only leaders) to maintain a sharp lookout for the symptoms of cortical inhibition and to always be prepared to take action and help the individual(s) snap out of it. In cases such as this one, where hearing loss is a factor, face-to-face communication is essential. **It can be as simple as walking up to the individual, looking them in the eye, and calmly saying out loud, "Take a couple of long, slow, deep breaths and tell me what's going on around you."** As the individual begins consciously deep breathing and speaking calmly they will engage their neocortex, which puts a lid on emotions and enables them to focus on what their senses are telling them.

I learned this technique from one of my Ranger battalion commanders, who taught it to me after I got overly excited during a training exercise. and screamed over the radio, "We found the precious cargo." I've been deep breathing before speaking on the radio ever since. I've used the same technique to help fellow warriors "snap out of it" during chaotic combat moments in Panama, Bosnia, Afghanistan, and Iraq. Engaging our neocortex and vanquishing anger, fear, and panic isn't voodoo magic. It's how our brains are hard-wired to make sense of the world around us and sensible choices about what to do next.

Like a coiled spring releasing, when Squad Leader #2 screamed over the radio, "Let's get the fuck out of here," the driver of the lead Humvee stepped on the gas and sped around the side of the Jinga truck while three of the five gunners in the back swung their weapons around toward the northern high ground. After passing a 200-meter-long rocky ridge to their right (west), the Canyon Road opened up on both sides, which enabled the Rangers in the lead Humvee to, for the first time, lay eyes on the spur. The evidence from all four investigations supports that "it was during this timeframe that the Afghan soldier was standing on the spur firing his AK-47 in the general direction of, and directly over, the section of the road that the lead Humvee was now driving on."

"Contact right!" one machine gunner remembered hearing as they drove out of ambush canyon. "It was like twilight," the Ranger recalled. "You couldn't see colors, but you could see silhouettes," another said.

6:45 PM., Lead Humvee Machine Gunner: "When we came around the final corner, it was dusk, and when I looked up on the spur, I saw the shape of a man holding an AK-47 that was spitting fire at us. The ride was bumpy as hell, so we couldn't focus our eyes on any one specific spot. Someone yelled 'Contact right,' and I wasn't sure who said it."

Squad Leader #2, front seat, lead Humvee: "I noticed muzzle flashes coming from a spur east of the village, and all of us started to shoot towards the muzzle flashes."

Ranger machine gun teams are trained to work together based on commands from their squad leader—in this case, Squad Leader #2. The squad leader controls the machine gun teams by communicating with standardized verbal commands and hand and arm signals. If the machine gun teams can't hear or see the verbal commands and hand and arm signals, they are trained to shoot where their squad leader shoots.

According to the first investigation, "several Rangers in the back also saw muzzle flashes and immediately started to engage." The muzzle flashes they saw were coming from the Afghan and his AK-47, as well as other members of Serial #1 who, according to

*Light level as the lead Humvee from
Serial #2 exited the canyon (approx. 6:45 P.M.).*

investigators, were "still trying to place fire on both the southern and northern high ground."

"The machine gunners saw only shapes," an investigator wrote, and all of them "directed bursts of machine gun fire without positively identifying the shapes."

Ranger Private: Before I joined the military, I believed in God and was pretty religious. During the first engagement that day, I was laying behind the boulders watching tracer rounds whiz by overhead, and I got a little panicky and yelled, 'God, can you get us out of this?' and Pat, who was laying behind his boulder, yelled back, 'Snap out of it, just keep your head in the game and focus.' And he was right. When I focused on where the bullets were coming from, it helped me to figure out how to stay behind cover so they wouldn't hit me. Pat didn't say it because he was an atheist, he said it because he cared about me. I still believe in God, but Pat was the one who kept me from getting shot that night."

With the exception of the Afghan, the rest of the Rangers on the spur were lying down behind cover when the lead Humvee came around the corner, which made them extremely difficult to

Serial #2 exits the canyon

When Serial #2 first laid eyes on the spur, the Afghan was standing up and shooting his AK-47 over their heads to the south, while some of the Rangers behind the Afghan were still engaging targets on the southern and northern ridges. (The light level in this picture is 12:00 P.M.; the firefight happened at approximately 6:45 P.M.) For best perspective, see pictures in color at firefight.commonsenseway.com

see. If you look inside the oval in the picture above, you will notice that the individual role players are all standing up and are still difficult to see.

Below is the same picture at a 254% zoom, and we can better see the shapes of individual role players standing in the open on top of the spur. We can also better appreciate the natural camouflage provided by the rocks and foliage around them. It took me a few seconds of searching to find the soldier who is standing in front of the boulders (just below and to the left of the cluster of three men). Once again, this picture was taken at noon with the sun at full illumination instead of 6:45 P.M., when it was setting behind the far mountain.

In stark contrast, the view from Serial #1's position on the spur was unobstructed and clear. (See picture below.) The creek bed

Close-up of men on the spur and the natural camouflage provided by the rocks and foliage.

View of the Humvees in the creek bed from the vantage point of the boulders on the spur. Note how easy it is to see the men and vehicles in the creek bed.

is mostly one color—sandy brown—and its texture is smooth, so instead of camouflaging the Humvees and making them harder to focus on, the smooth, monochromatic creek bed made the vehicles easy to see, even in low light.

The men of Serial #1 arrived at their positions on the spur from the west (see map of their route of travel), so they had no idea what the spur looked like from the creek bed below. **As a result, the unobstructed view of the Humvees from the spur may have led the men of Serial #1 to believe that Serial #2 could see them just as clearly as they could see the Humvees of Serial #2.**

I asked the Ranger private who was with Corporal Tillman behind the boulders what he thought when he saw the pictures for the first time, nineteen years after the firefight.

Ranger Private: "I had a lot of bitterness toward those guys after the fire-fight because I couldn't reconcile how they could not have seen us. They kept saying the same thing over and over—'We couldn't see you,' or 'We never saw you'—and no one believed them. I know it was dark, and it's really hard to see anything when you're riding in the back of a Humvee, so I finally figured they had to be telling the truth. It helped me get over the bitterness. When I saw the pictures, it reinforced what they were saying from the beginning. I wish they would have let us see those pictures a long time ago; it would have helped with closure."

Although the creek bed in front of the spur was less confined than inside the canyon, the ground they were driving over was just as rocky and equally undulating, making the effect on the driver the same. To find a path of least resistance, the driver had to steer the vehicle around piles of river rocks, boulders, and one- to two-foot-deep pools of water. The effect of the zigzagging, bumpy ride on the machine gunners and their weapons was also the same. The bumpy ride made it difficult to focus their eyes on any one specific spot. While the zigzagging caused massive swings in the horizontal and vertical movement of their barrels, making it difficult to control the cones of fire and beaten zones of their bullets.

In addition to Squad Leader #2's M4, the Ranger machine gunners in the back of the lead Humvee were engaging the spur with

View of Corporal Tillman's position behind the boulders from the rock wall corner. Distance is approximately seventy-five meters/eighty-three yards.

the M2 .50 caliber machine gun, the M240 7.62 machine gun, and the M249 5.56 machine gun.

According to the first investigation, this was when some of the Rangers on top of the spur started realizing, "Hey, we're being fired on from our own guys." They could tell by "the sounds of the weapons being fired and color of the tracer rounds" as they snapped and cracked while whizzing overhead. Most of the Rangers on top of the spur began low-crawling backwards to maximize the protection provided by the military crest. Due to their position behind the boulders on the front-facing portion of the spur, backing up for further protection was something that Corporal Tillman and the Ranger Private beside him no longer had the option to do.

Ranger Private: "That's when Pat also realized, 'Hey, these are our guys shooting; they are friendlies and they're firing at our position.'"

To mark his position as friendly, Corporal Tillman threw a smoke grenade. Only two Rangers interviewed by investigators—the

Light level when smoke grenade was thrown, as seen from the corner of the rock wall at the same time of day, and same time of year (6:48 P.M., 22 April) as the firefight. The photo was taken during a reenactment conducted for the fourth investigation on 22 April 2006.

Ranger private next to Corporal Tillman, and the platoon sergeant who was riding in the back of Serial #2, said they actually saw the smoke. Whether it was too dark to see the violet-colored smoke, or the smoke grenade simply malfunctioned, no one could say.

On 22 April 2006, two years to the day after the firefight happened, the fourth investigation went to the actual site at the base of the spur and conducted a videotaped reenactment at the exact time of day that the smoke grenade was thrown. (See pic below.) In the low light, it's extremely difficult to see the smoke, but you can clearly see the magnesium fire and sparks from the grenade as it burns the chemicals that make the smoke, which could easily be misidentified as muzzle flashes or ammunition cook-off.

A few seconds after Corporal Tillman threw the smoke grenade, the Ranger machine gunners stopped firing. According to statements given to investigators, the Ranger machine gunners weren't holding their fire because they saw the smoke; instead, they were holding their fire because they could no longer see the top of the spur.

It wasn't the smoke that was blocking the machine gunners' view. The view of the spur was blocked by a combination of a dip in the creek bed and a six-foot-high rock wall they were driving parallel to. (See photo below.) As a result, the machine gunners had to hold their fire until the vehicle reached a position where there were no longer any obstacles between them and their targets. Time and distance estimates made with a similar Humvee during the fourth investigation concluded that the vehicle was traveling at a speed between ten and fifteen miles per hour when it made the turn, during which time the rock wall blocked their view for approximately four to five seconds.

Ranger Private: "We were yelling, 'Stop, stop, friendlies, cease-fire,' but they couldn't hear us. They stopped the friendly contact for a few moments, and we thought the battle was over, so we were relieved. We both stood up, and it was at this time that their vehicle rolled back into view in a better position to fire on top of us; it was actually perpendicular now. Then they started firing again. I heard an M4 fire first, then the .50 cal. opened up again."

Humvee driving past the corner of the six-foot-high rock wall.

Note: When I first read this statement by the Ranger private, I mouthed the words: "Get down." My reaction was one more example of the relationship between experience in combat and experience with friendly fire. All I could do was wish I had been there with them on the spur that day. That thought brought me back to Sergeant Major (ret.) Ted Kennedy's wise words:

> *"Remember when we were just starting out back in the late 80s and 90s? Guys like us read everything we could get our hands on about lessons learned from past combat operations…We have an ongoing responsibility to future leaders to ensure they learn from situations like this…"*

According to investigators, none of the Rangers manning machine guns in the back of the GMV identified friendlies on the spur either the first time it came into view or the second time after they passed the six-foot-high wall.

The sun had already set behind the mountain, and the combination of low light and the spur's mottled, rock and foliage-covered surface made shapes and details difficult to distinguish. While it was definitely dark enough to impair the ability to see, it was not yet dark enough to use night observation devices (NODs).

The driver of the lead Humvee only identified friendlies on top of the spur after the vehicle rounded the corner of the rock wall and after he had seen their vehicles first. Investigators later confirmed that the rock wall was 72 inches high and the driver's eye height was 62 inches off the ground, which meant it wasn't possible for him to see Serial #1's vehicles before his GMV rounded the corner of the rock wall.

And it wasn't just the lead Humvee in Serial #2 that had a difficult time making sense of what they were seeing up on the spur. The vehicle behind the lead Humvee was a cargo Humvee carrying members of the mortar section and a couple of snipers, all of whom had fired their weapons at enemy positions on the northern high ground from inside the canyon. According to the Ranger private next to Corporal Tillman: "The cargo Humvee came down the road toward our direction; when they made eye contact with us, they opened fire with small arms."

Here's how the Ranger in charge of the cargo Humvee described what happened after they broke free from the canyon: "After we passed the Jinga truck, there was a one-hundred-meter gap between us and the lead vehicle [Squad Leader #2's Humvee]. That vehicle continued to fire the .50 cal as they moved. When we rounded the last curve, I saw them stop again and began shooting at the spur short of the village. We stopped and dismounted from the vehicle, and the snipers [who were armed with SR-25s that fire 7.62 caliber bullets] took shots, and then we started moving again…I didn't even know there were personnel up there until I got parallel with the spur [after turning the corner of the rock

Once the Humvee was past the corner, the boulders (see oval) provided less protection. note stone house on far left where platoon leader and radio operator are located.

wall], and then I looked up there and could see the friendly troops on the ridge. **I did not see them there until I was right by them.**"

In addition to low light conditions, camouflage, temporary hearing loss, and stress, what other factors contributed to the difficulty that the members of Serial #2 were having in "making sense" of what they were seeing on the spur?

Once a firefight starts, there is typically much dust, smoke, small-arms fire, explosions, yelling, and sensory overload. Confusion reigns supreme, making identification of friend or foe an ongoing challenge. In situations like this one, where two or more elements are unsure of each other's exact location, there are some common factors that strongly influence the way our brains make sense of what we perceive. One of the most important factors is called "location assumptions."[80]

Our brains make location assumptions in order to bridge the gap between what we know and don't know about the location of the things around us. What we know about location of "things" comes from two primary sources: 1) our short-term memory of where things such as our car, our keys, or our friends were located the last time we saw them, and 2) our foundational knowledge of everything we know about those things (e.g., cars, keys, friends' behavior patterns) stored as long-term memory.

As an example, imagine that you just entered a room you've never stepped foot in before, and suddenly the lights go out and the room goes pitch black. Even though you were only in the room for a few seconds before the lights went out, your short-term memory can easily recall the last most updated location of the things around you (e.g., where the door is that you just walked through, the location of the coffee table and chairs off to your left, the location of the wall off to your right, etc.).

Your brain combines this knowledge from your short-term memory with all the knowledge you know about these things stored in your long-term memory (e.g., *watch out for that coffee table; the wall will guide me back to the door; there should be a door knob approximately four feet above the floor and a light switch on the knob side of the door*). To make sense of where you are and what to do next, your brain begins combining and connecting this information to create location assumptions.

It works the same way in a firefight. As Vietnam veteran John T. Reed explains it: "Let's say that when the firefight begins, the good guys are on the south side of a road and the bad guys are on the north side. It would be imprudent for any of the good guys to cross that road to the north side without first making sure that all the other good guys know exactly where they are crossing and where they're going after they cross."[81] The same would apply to other prominent location borders like water versus land, forest versus clearing, and high ground versus low. Conscious awareness of the way our brains make location assumptions enables us to see the world through the mind's eye of others and to anticipate the future as it unfolds in front of us.

When asked by investigators about Serial #1's location, most of the Rangers in Serial #2 said the same thing as the regimental sergeant major: *"I didn't have any idea when we first made contact that Serial #1 was even close."* **When the Rangers in Serial #2 broke free from the canyon, everything they knew about the location of Serial #1 was based on location assumptions.**

The most recent memory they had of Serial #1 was of them driving out of Magara in their Humvees and Toyota Hiluxes on their way to Mana. They assumed, based on their recent experiences conducting missions together, that if Serial #1 was ambushed by an enemy force located in the mountains, they would follow their platoon standard operating procedures and do what Serial #2 did: return fire using the crew-served weapons mounted on their vehicles, from which they could suppress and destroy the enemy, or drive out of the kill zone to a position where they could. **As a result, they assumed that the next time they saw their teammates in Serial #1, they would be waiting in and around their vehicles, near the town of Mana.**

With regards to the enemy, Serial #2's **experiences over the prior eight to ten minutes reinforced that the enemy was shooting at them from the northern high ground.** With regards to the terrain around them, their experiences over the last eight days taught them the same thing the foreign fighters said in the Asia Times article: *"One crosses the first mountain and sees a similar mountain emerge, and after crossing another mountain, he feels a spin in his head and thinks the whole world in this area is the same and leads the way nowhere."*

Thus, when the first two vehicles of Serial #2 finally broke free from the confines of the canyon walls, they found themselves in a location they had never driven through before and had never had time or reason to study and familiarize themselves with on a map. When they saw the Afghan on the high ground to the north firing his AK-47, as well as at least one muzzle flash from the Rangers directly behind the Afghan on the spur, it reinforced their location assumptions about the enemy and the terrain. Based on everything they had experienced over the previous ten minutes, they assumed that what they were seeing and hearing on the northern high ground was the enemy, and that enemy was still trying to kill them.

Top of the Spur, Serial #1: One of the Rangers on top of the spur told investigators he observed, "the platoon leader and his RTO [radio operator] run for cover as bullets from the guns of the lead vehicle began impacting the corner of the stone house they were seeking cover behind."

Recall that the radio operator had earlier spotted an enemy position located on the northern high ground and fired "half a magazine" at them. He and the platoon leader therefore assumed

View of the stone house (two-story structure on left) from the lead Humvee. If you look closely, you can see role players standing at the location where the platoon leader and radio operator were injured.

that the enemy was on the northern high ground, and the safest location for them to seek cover from the confusing crossfire was on the southern side of the wall and stone house. But instead of making them safer, this position completely exposed them to the crossfire of friendly forces and also inadvertently provided the confused machine gunners of the lead Humvee with one more moving target to mistake as an enemy. It was at this point that the radio operator was shot in the knee.

The dilemma of location assumptions is not actually in making them; rather, it's in the basis or foundation of knowledge upon which they stand. Assuming the chairs in a blacked-out room won't reorganize themselves after the lights go out is a safe assumption based on a solid foundation of knowledge (e.g., inanimate objects don't move on their own). Location assumptions are reliable when dealing with inanimate objects such as furniture because their location is predictable. The same cannot be said when dealing with animate objects such as humans. Humans do things that no one, including themselves, can predict.

View from the stone house looking east toward the slot canyon and the spur. The far Humvee is parked at the corner of the rock wall. The near Humvee is parked where the Rangers in the lead Humvee of Serial #2 took their final shots and ceased fire.

Never assume you know where another person or team is located unless you can see, hear, or touch them with your senses. If you can't, then you should always assume they are somewhere you don't expect them to be. Expect the unexpected.

One of the hallmarks of common sense is how implicitly aware we are of our assumptions. Recall that assumptions are the biological way our brains are hardwired to fill in the gaps between what we know and what we don't know about what's going on around us. Serial #1 left their vehicles and ran to the top of the spur because Squad Leader #1 and the platoon leader assumed the high ground would enable them to communicate with and provide fire support for Serial #2. This was a solid assumption based on a sensible purpose: to help their fellow Rangers.

This purpose—to help their fellow Rangers—made sense of their choice of paths—to leave the vehicles and run up to the top of the spur. However, it also underscores the adaptive nature of decision-making in combat.

1) Purpose + 2) What's Going On Around Us
= 3) Why We Choose What We Choose.

1 + 2 = 3 To Make Sense it has to Add Up

In moments of incomprehensible complexity such as a fire-fight, every path we choose is a life or death choice. Choosing is how we humans adapt. **When the situation going on around us changes, so too must our choice of paths. We must have freedom to adapt.** As it turned out, the spur didn't enable Serial #1 to communicate or provide fire support. Yet no one in Serial #1 was consciously aware that what they were doing on the spur was based on assumptions.

We can create conscious awareness of our assumptions by training our brains to say them out loud. Repetition breeds recognition. Once we say an assumption out loud (e.g., "We're going to run up to the high ground to see if we can provide fire support and make positive comms with Serial #2; if it doesn't work out, we'll head right back down."), everyone understands it's an assumption and can constantly pressure test its validity against the reality of the on-going situation.

As soon as the lead Humvee passed the corner of the rock wall, the driver told investigators: "I looked to my right and saw one pax [person] with an AK-47, which confused me for a split second, but then I saw the Humvees up ahead and the rest of Serial #1 on the top of the ridgeline. I yelled twice, 'We have friendlies on top.' They [the machine gunners in the back] must have not heard me because they continued to fire on them. I screamed 'no'; and then yelled repeatedly several times to ceasefire. No one heard me. **By that time, I believe everybody was deaf from all the gunfire that had been shot off."**

One reason the machine gunners resumed firing was because their squad leader, who was sitting in the front passenger seat, continued firing. Squad Leader #2 continued firing in the direction of the Afghan as soon as he came back into view.

Squad Leader #2: "I was still shooting and engaging what I now know was the Afghan soldier when I first heard the driver screaming friendlies. I looked down at the road, then back up the ridgeline, and I identified Rangers. That's when I yelled ceasefire. **I had tunnel vision,**" he later explained to investigators.

When the Humvee turned the corner and Squad Leader #2 saw the Afghan firing his AK-47 directly at or over his vehicle, he chose to defend himself and his men by firing back. However, by focusing his attention through the lens of his advanced combat optics gunsight (ACOG), he lost situational awareness of his men, their direction of fire, and every other thing that was going on around him at that moment in time.

It is unclear whether the Afghan soldier was killed before or after the lull in fires (he was hit by at least seven rounds). What is clear is that as soon as the lead vehicle drove past the corner of the six-foot-high rock wall, the Rangers manning the machine guns mimicked their squad leader by continuing to fire their weapons in the same general area where the Afghan had just stood.

Critically, this was also the same general area (ten feet further up the spur) from which Corporal Tillman and the Ranger private were trying to signal them to stop. Corporal Tillman and the Ranger private next to him weren't the only ones trying to signal the Rangers in the lead Humvee to stop firing. The Rangers on top of the spur were also waving their hands and yelling ceasefire, however at least one of them continued to fire his weapon at suspected enemy positions on the high ground around them.

Top of Spur, Serial #1 Ranger: "As I moved down the ridge to spread out, I could see and hear rounds coming at us. These rounds were tracers, and I could see and hear Tillman and the Ranger next to him yelling, 'Stop firing.' I then moved down the hill to push out a little further to the right. I then started firing on muzzle flashes from the hillside to the front of us."

Top of Spur, Squad Leader #1: "I noticed the first vehicle, which stopped, and I saw the M240B gunner in the back of the first vehicle fire a burst towards me. The burst walked its way past my position to the north. I then set off the pen gun flare and began motioning ceasefire by waving my hand and arm over the front of my face, palm out.[82] There were other bursts, but not as close. Just then, somebody fired upon Corporal Tillman's position because I saw dirt and rocks being kicked up at his location. I then heard people in the vehicle yelling ceasefire."

Ranger Private: "I myself had taken cover by the rock. Pat taking cover to the side, and the AMF soldier was just pretty much standing out in the open, shooting up at the ridgeline. Not too long after, we took a couple pot shots from a gun-mounted Humvee that was heading towards us. Nothing serious, Sir. Maybe a couple bursts of an M4. I kind of got down a little bit and tried, you know, to assess the situation, what was going on, because I was actually positive we were being shot at at that time."

Investigator: "And when was the AMF soldier hit?"

Ranger Private: "He wasn't hit until the second encounter of friendly fire, Sir."

Investigator: "What about Corporal Tillman?"

Ranger Private: "Pat was behind some pretty good cover, to where he wasn't really in too much danger, and I was completely in the open for getting shot. I was watching the rounds where they were going, and Pat could see and look around—and I was noticing that most of their fire seemed to be directed towards me. The AMF guy, he was dead at the time. He was lying down. I could see him lying down, and I realized that they were predominately shooting at me, and I guess Pat did, too. And he moved out from behind his cover to throw some smoke or whatever he did, because I initially thought it was a pen gun flare. All I remember him telling me was, 'Hey, don't worry, I've got something that can help us.'

View of Corporal Tillman's position (boulder on left). The distance from the boulders to the vehicle is seventy-five meters or eighty-three yards.

And he popped a smoke, I guess, and that's when he got shot—one of the few times."

Investigator: "When did they stop shooting at you?"

Ranger Private: "They stopped shooting at us not too long after Pat threw the smoke. I just remember him throwing the smoke, and then he started having a cry in his call, you know, and he started screaming, 'My name is Pat Tillman,' and he said that probably five to ten times, and then he went silent completely, and then maybe five seconds after that—I mean, not too much longer, maybe fifteen seconds they stopped shooting. **I wouldn't have lived if he had done nothing. The angle from where they were firing from would have killed me, not Pat. If he had done nothing, he would be alive, and I would be dead. Pat gave his life to save mine.**"

The machine gunners in the lead GMV finally ceased firing as they lurched to a stop near the 2-story stone house. An AMF soldier wearing tiger-striped BDU's ran up to their vehicle while frantically waving his hands and yelling, "stop, no, stop" over and over. Deaf from all the gunfire in the canyon, dumbfounded by their location assumptions, and blinded by cortisol-induced frustration and stress, the Rangers in the lead Humvee didn't keep shooting because someone told them to. **They kept shooting because their stressed-out, cortically-inhibited brains couldn't make sense of what was going on around them.**

Top of Spur, Squad Leader #1: "After hearing the ceasefire called, I heard the Ranger who was with Corporal Tillman freaking out. I crested the spur and saw the Ranger without his helmet and weapon grabbing hold of Corporal Tillman's shoulder. He yelled at me, 'It was our own guys that did it.' Corporal Tillman was in the sitting position between the rocks, with his legs facing the road and body slumped over. Then I saw the AMF soldier laying approximately ten feet in front of Corporal Tillman heading down towards the canyon. I then called over the platoon sergeant to advise him we had two eagles[83] KIA."

6:48 P.M.: The total engagement time from the initial contact made by Serial #2 in the canyon and the final shots fired by the lead Humvee was no more than fourteen minutes. The evidence would later confirm that Corporal Patrick Tillman died as the result of **friendly fire** during the final seconds of the firefight. Forensic pathology would reveal that his head wounds resulted from M249 5.56 rounds, and the rounds that hit his chest were either 7.62mm rounds or 5.56 rounds, or both.

6:50 P.M.: Sixteen minutes after Serial #2 had entered the canyon, 2nd Platoon reported that its forces "were no longer in contact." As a Ranger-appointed investigator later put it, it was unclear both then and later who the Afghan attackers spotted by half a dozen Rangers in both serials had been, how many guerrillas there were, or whether any were killed.

6:52 P.M., Reconnaissance Platoon Sergeant (Sergeant Rex): "We heard the last shots fired as we were pulling into Mana. The first guy I saw was the weapons squad leader, who couldn't hear a thing and was screaming at the top of his lungs, 'What happened?' The next guy I saw was [Squad Leader #2] and he was bleeding from both ears, which usually means your eardrums are blown out, so I didn't even try to talk to him. Then I found the platoon leader and he had a mouthful of gauze—it looked like he had been shot in the face—and when he saw me, he spit out the gauze and said, 'How did you get here so fast? I am so glad to see you.' When I bumped into the platoon sergeant, we both looked at each other and said, 'What the fuck are you doing here?' The entire platoon was an emotional wreck; only later did I find out about all the insanity they had been through earlier that day.

"While the 2nd Platoon was busy consolidating and calling in Medevac helicopters, me and a couple of my guys walked back into the ambush canyon. All I can say is those guys were lucky they all weren't killed. No one should have come out of that canyon alive. It would have been a different story if they had comms between them. If the two serials from 2nd Platoon had known we were out there, they could have called us on the radio, and we could have relayed messages between them."

I asked Sergeant Rex for his thoughts on the enemy and what he believed happened to them.

Sergeant Rex: "Me and my guys talked about it, and we started asking each other why the entire village of Mana was on their roofs right before the ambush was initiated? Were they on their roofs to watch the ambush, and then we distracted them? Maybe the enemy thought we were the reinforcements and they called off the ambush because they figured we were about to flank them. We'll never know whether our arrival had any effect on them, but I do believe that the platoon's suppressive fires played a role in driving off the enemy. Whenever the enemy realizes they're outgunned, they tend to fade away so they can fight again another day."

At least five members of Serial #2 and six members of Serial #1 told investigators they positively identified enemy personnel

on the northern high ground. While doing research for this book, I learned there was an Afghan Special Forces Unit operating in the same general area (Spera District, Khowst Province) that night. These Afghans were outfitted with state-of-the-art equipment, were trained to a higher level than any other Afghan military unit, and were supervised by private contractors from other U.S. Government agencies. Many of these men fought valiantly against the Taliban and foreign fighters until they were abandoned in August of 2021.

On this night, the Afghan Special Forces Unit was operating independently, so my source had no knowledge of their exact location during the firefight. Only that they were in the general area to kill or capture enemy insurgents. Although none of the four investigations mention anything about Afghan Special Forces Units or their purpose that night, the Asia Times Online article (page 90) provides a potential clue based on their role in Operation Mountain Storm:

> *By occupying the area, the U.S. hopes to deprive the insurgents of the tribes' crucial support. **Forced to flee, the insurgents would eventually fall into the hands of the United States' local proxy networks of anti-Taliban tribes and warlords.** Such is the plan."*

This area is also crawling with Afghan and Pakistani smugglers and bandits *(they were dressed in really weird clothes, like they were from somewhere else)*, so there are lots of possibilities regarding who opened fire on Serial #2 and whether or not those that did were local tribe members, confused Afghan Special Forces, or the insurgent enemy the Afghans and Rangers were pursuing.

6:53 P.M.: The rest of Serial #2 arrived in the vicinity of the rock wall, as the squad leaders began checking on their people to ensure they weren't injured and that all of their weapons and

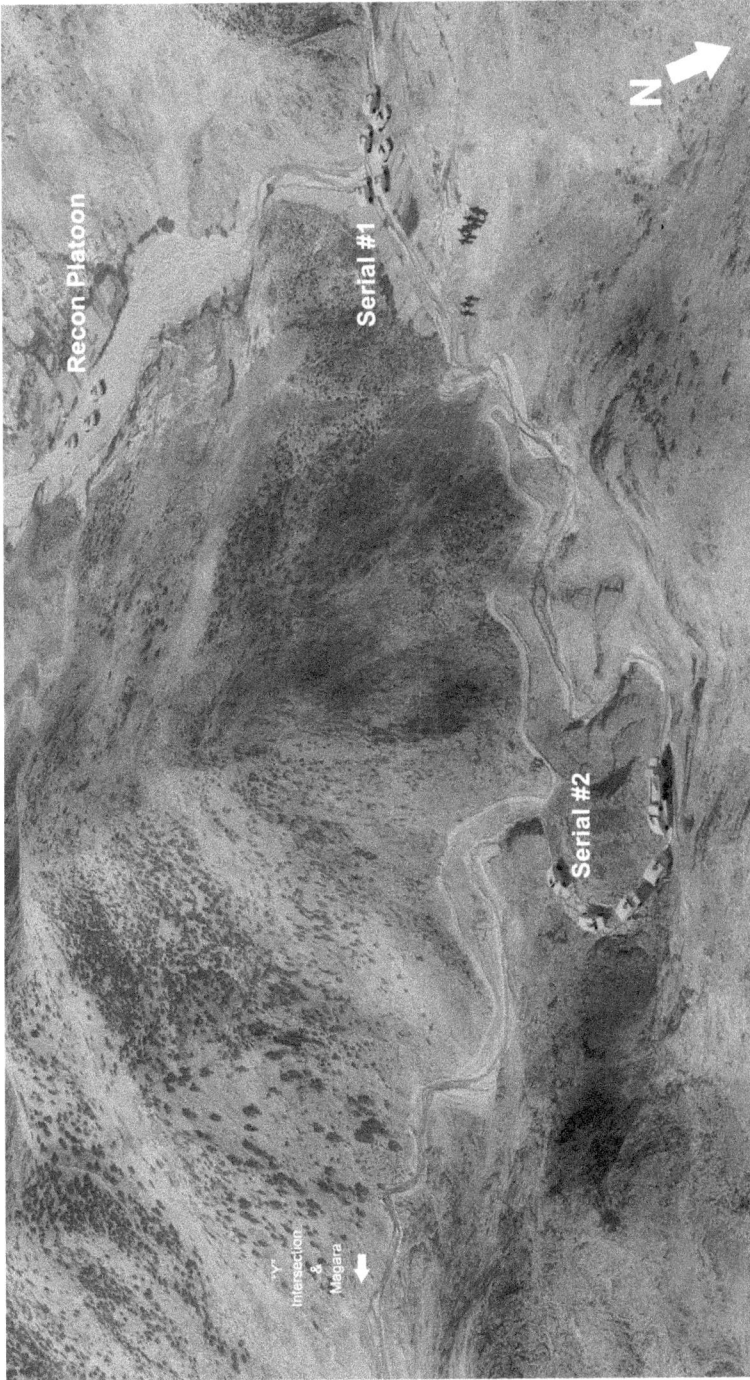

View from the northern high ground reveals enemy fighters were able to see and shoot at all three Ranger forces as they converged on Mana. The northern high ground was key terrain for controlling this area. This view confirms that the enemy could see the Reconnaissance platoon as they approached Mana from the south, and may have believed they were reinforcements.

equipment were accounted for. Kevin Tillman's vehicle was the last to pull into the perimeter. **"You okay Tillman?"** asked his organic squad leader (who had been riding in the Jinga truck with Pat's organic squad leader). "Roger that, sergeant, but this gun is all f----d up. I wasn't able to fire it in the canyon cuz I couldn't get it to charge. Probably okay anyhow, cause I don't think I could have fired it in the canyon. I got off a few rounds with my pistol," he explained. "Okay," replied his squad leader, "I'll take a look at it in a minute, right now I need you to stay here and pull security while I go find the platoon sergeant and see what's going on."

The platoon sergeant was standing on the spur with Squad Leader #1, ten feet in front of the boulders. "He's dead," said Squad Leader #1. The platoon sergeant just stood there staring in silence. When I asked him if he'd share with me what he was feeling, he responded with one word: "devastation." He still chokes up when he talks about it, as do the other sergeants who gathered during those final moments before darkness to pay their respects to a fallen comrade and discuss what the platoon should do next.

As word of what had happened spread throughout the platoon, the feeling of devastation spread along with it. One common denominator that jumps out when you read all the statements is that Pat Tillman wasn't just liked and respected by his teammates; he was revered.

Unable to reach the platoon leader on the radio and frustrated that no one seemed to know what had just happened, the platoon sergeant searched the area and found him sitting near the stone house, propped up against a rock wall.

"At that point, I spotted the blood around his mouth," the platoon sergeant later told investigators. The swelling from his wounds drastically inhibited his ability to communicate, and the platoon sergeant instantly recognized he was about to go into shock. The radio operator was still talking on the radio and attempting to coordinate Medevac helicopters without realizing he'd been shot in the knee and hit by mortar or grenade shrapnel in the neck and chest.

7:23 P.M.: Darkness had now completely set in, making it difficult to see anything without night vision goggles, which Kevin Tillman was wearing as he watched four Rangers struggle to carry a body bag down from the spur.

According to statements Kevin Tillman provided to investigators: "I asked, 'Who's that?' and one of the Rangers said, 'It's an AMF soldier,' and that didn't make sense, so I started to get nervous because my brother is a pretty loud type guy, you know, and then they decided to move my vehicle up by the other vehicles. I was standing in my turret looking through my night vision goggles, and I still didn't hear anything, and I got a weird feeling, so I asked [redacted], 'Where's Pat?' and he said, 'I don't want to be the bearer of bad news, but he's KIA.' So forty-five minutes after the firefight, I find out Pat's dead, and the only reason I find out is because I ask?"

Note: *I asked the platoon sergeant about notifying Kevin, and here's what he said:* "We were all devastated by Pat's death, but we were still in a battle, and we had no idea whether the enemy had run away or was trying to flank us. We also had two critically injured Rangers who needed an immediate Medevac. Our priorities were making sure the area around us was secure, getting the wounded and dead down to the helicopter landing zone, and setting up the landing zone so we could guide in the Medevac helicopters. Once we had everything setup, one of the squad leaders told me that Kevin already knew."

7:35 P.M.: Two UH-60 Blackhawk Medevac helicopters landed in the vicinity of the rock wall.

7:47–7:56 P.M.: The injured platoon leader and his RTO were loaded onto one Medevac helicopter, and the bodies of Corporal Tillman and the Afghan soldier were loaded onto the other. The

Medevac helicopters lifted off and flew directly to the 325th Field Hospital located at the forward operating base in Khowst. Corporal Tillman was officially pronounced dead at 7:50 P.M..

8:20 P.M.: Kevin Tillman was flown on a CH-47 helicopter from the ambush site near Mana back to the FOB in Khowst.

8:30 P.M.: Kevin Tillman arrived at the FOB in Khowst. The battalion operations officer/S-3 was the first person he talked to.

The operations officer/S-3 later told investigators: "Literally, I was the first person who talked to him after laying the flag on his brother, and I, as the acting commander, I said, 'Hey, I can't imagine what you're going through, blah, blah, blah.' And he said, 'Are we going to get them, Sir?' or something to that effect.

7:30 P.M. to 11:30 P.M.: Upon notification of what had transpired, the A Company commander, A Company first sergeant, and the entire 3rd Platoon from A Company immediately loaded their vehicles and drove the forty-four miles from the FOB in Khowst to Mana in order to assist and reinforce the platoon. As soon as they arrived, they began patrolling the area in and around the ambush site while relieving the men of the 2nd Platoon so they could get some much-needed rest.

Most Rangers told investigators they simply dropped to the ground around the vehicles and closed their eyes, but barely slept. One Ranger told me, "I was working on one of our radios, and my legs gave out and I collapsed. My buddy ran over and said, 'Were you shot?' And I actually didn't know what hit me until the medic told me I was going through an adrenaline crash."

Although he was less than three miles from the ambush site and monitoring the situation as it unfolded on his satellite radio, the battalion commander chose to stay where he was. As he later explained to investigators: "I was aware, as it was happening, of the fight going on and the events that transpired, as much as you could garner over the radio. So I knew, as it was happening, what was going on. **As soon as the sun came up and it was pretty safe to travel,** we got a convoy together and went over to the ambush site and linked up."

23 April, 9:30 A.M.: According to investigators, the battalion commander arrived on the scene between 9:00 and 10:00 am, almost fourteen hours after the firefight ended, and just as the 2nd Platoon was searching houses and "clearing" Mana. The platoon sergeant, who had been up all night trying to figure out what had happened, was the first one to talk with him.

Platoon Sergeant: "The battalion commander got out of his vehicle and walked up to me and asked, 'What happened?' I told him, 'We had no communications between the serials and didn't know where each other were.' I also told him that 'as Serial #2 was coming through the canyon, we took mortar and RPG fire, and it looks like Serial #1 got out of their vehicles and ran up to the high ground to provide support. The firefight was intense, and Corporal Tillman was caught in the crossfire. At this point we can't tell whether he was shot by friendlies or enemy but after piecing everything together, I'm pretty sure that it was fratricide.' He was just staring at the spur and boulders where Corporal Tillman was killed. Then he finally looked at me and said: 'He was a hero. He charged uphill to seize the tactical high ground from the enemy.' Then he walked away to talk to someone else."

Battalion Commander: "After talking to everyone, it was clear to me that the circumstances were there where a fratricide could have happened, but we had, as you know, probably from talking now—you know, you had enemy here, enemy here, and enemy over here [pointing to the investigator's map of the area]. And so they had been caught in this crossfire, and so I wasn't—while I was sure that—in my mind, I was sure that he had been shot at, I wasn't quite certain who had killed him because we had guys wounded up here in the city too. And so I said, 'Alright, I think we need to do an investigation.' So I called the regimental commander [located at Bagram Airfield, one hundred miles north] and told him my gut feeling was that Tillman had been killed by friendly fire. That was around noon on April 23rd."

While the platoon was searching houses in Mana, the platoon sergeant and three or four other Rangers who were pulling security on the perimeter heard helicopters approaching. When they looked up, they saw two CH-47 Chinook helicopters flying overhead. Attached to their bellies and hanging beneath them from cargo slings were two up-armored Humvees being delivered to the forward operating base in Khowst.

"It was a quiet reminder," one of the Rangers later told investigators, "that perhaps if our leadership had done their job and had gotten that helicopter to us like we asked, none of this would have happened."

From the moment the Ranger private next to Pat Tillman began screaming, "He's dead," and, "It was our own guys who killed him," everyone in the platoon began trying to piece together what had just happened. At least sixteen members of the platoon witnessed the final moments of chaos as the lead Humvee pounded the spur with interlocking volleys of machine gun fire. None of the sixteen Rangers had any doubt that Corporal Tillman was caught in the middle of the platoon's crossfire. The only doubt that remained was whether the bullet(s) that actually killed Corporal Tillman were fired from the weapon of a friend or the weapon of a foe.

Adding to the confusion was the fact that both the platoon leader and his radio operator were convinced that they were hit by enemy fire. Yet with every passing hour at the ambush site, evidence that the fire that killed Corporal Tillman was friendly continued to accumulate: there were .50 cal bullets lodged in the boulders around him; he was hit in the chest and forehead while facing south toward the road; and the Ranger who was inches away from him when he died was still telling anyone who would listen that he saw the lead Humvee shoot the bullets that killed Corporal Tillman.

All five key leaders on the ground at the ambush site (the platoon sergeant, the company commander, the company first sergeant, the regimental command sergeant major, and the battalion commander) told the investigators under oath that they concluded the fire that killed Pat Tillman was likely from friendlies, before they left the ambush site the next day (23 April).

Lead Humvee Machine Gunner: When the sun came up on the twenty-third, we began clearing Mana and talking with the locals. It was a long, hot, frustrating day. At dusk, almost exactly twenty-four hours after the firefight, we began loading our vehicles for the drive back to Khowst. As we sat in our vehicles, I watched as the platoon sergeant talked to the squad leaders off to the side. When their meeting ended, our squad leader walked over to us, and he didn't look happy. The cruel truth began to be spoken. Pat wasn't killed by the enemy. He was killed by friendly fire. They didn't have any autopsy results yet, but there were .50 cal rounds lodged in the boulders in front of Corporal Tillman's position. **The immediate implication was that the .50 cal gunner fired the rounds that killed Corporal Tillman. But that's now what I was thinking. I was thinking that I fired in the same places as the .50 cal. It could have been any of us."**

The Ranger regimental commander immediately put out guidance stating, "Until the investigation is complete, until we know what happened, I do not want communication of the ongoing investigation going outside the unit." This was not unusual guidance for special operations units in combat zones. What was unusual was the way the guidance was implemented by the Ranger chain of command.

23 April, 11:00 A.M.: Hearing that the platoon leader and his radio operator were being treated at the Khowst field hospital,

Kevin Tillman walked over to pay them a visit. They talked briefly about what each of them saw, heard, and understood had happened. Because they were immediately separated from the rest of the platoon after the ambush, these were likely the only three Rangers from 2nd Platoon that didn't know or highly suspect that Corporal Tillman was killed by friendly fire at that time.

Later that afternoon, all three men attended a memorial service, during which every Ranger at the base had the opportunity to pay their final tributes and farewells to Corporal Tillman. The memorial service ended with "Taps" playing in the background as the flag-draped casket was loaded onto a CH-47 helicopter in preparation for a forty-minute flight to Bagram airfield.

As the honor guard loaded his brother's coffin, Kevin Tillman spoke with the battalion commander and the operations officer/S3. According to investigators, Kevin asked that they promise to do everything possible "to find the guys that killed Pat," and everything they could "to find Pat's journal." They told Kevin they would "do everything they could to find the guys" and "they'd leave no stone unturned trying to find Pat's journal."

And then they chose to tell him nothing else.

"Let us know if you need anything," were the final awkward words spoken to Kevin as he walked across the tarmac and boarded the CH-47 to accompany the body of his brother on the helicopter flight back to Bagram Air Base.

In the context of the moment, while standing on the airstrip in Khowst, those leaders made a foundational choice. By choosing not to tell Kevin Tillman the truth about what they already knew, the chain of command was choosing to lie to him instead. By proxy, they laid the foundation upon which all of the lies and half-truths that followed would be built. The only way to build on a foundation of lies is with more of them. And for the next thirty-four days, the chain of command would go to extraordinary lengths to do just that.

Freeze this moment in time on the airfield and put yourself inside the situation. Every option was still open to the chain of command at this time. This was the first opportunity the leaders of the battalion had to share the facts as they knew them with Kevin Tillman by simply saying out loud what everyone else up and down the chain of command already knew at that moment in time.

One of the primary purposes of Investigations #3 and #4 was to determine whether there was a concerted effort by the leaders above the platoon level to cover up the fact that Pat Tillman was killed by friendly fire.

Investigator #3: "Tell me what happened after you arrived at the ambush site?"

Battalion Commander: "I did not tell Pat—or, not Pat— his brother. He left on the night of the twenty-third, I think the next night, I had the opportunity to explain to him what we were doing. And I told everybody—I said, 'Don't tell him yet. Let's wait and keep this between us until we get more facts because we don't know, and he's got to meet with the guys in this unit.' So to that degree, I probably could have been more forthright with him. But I talked to the regimental commander [his boss] about it earlier, and we had agreed that it was the best way."

The last Ranger to see Pat Tillman alive was the Ranger private crouched next to him behind the boulders. According to his sworn testimony provided during congressional hearings, he was ordered by his chain of command "not to divulge," especially to the Tillman family, "that friendly fire killed Pat Tillman."

Ranger Private: "I wanted right off the bat to let the family know what had happened, especially Kevin, because I worked with him in the platoon and I knew that he and the family all needed to know what had happened. I was quite appalled that **when I was actually able to speak with Kevin for the first time on the phone two days after the ambush, I was ordered not to tell him**

anything." When asked who gave him the order, he replied: "It came from my battalion commander; he basically just said, 'Do not let Kevin know,' and he made it known I would get in trouble if I spoke with Kevin on it being fratricide."

Kevin Tillman wasn't some random soldier from another unit. He was a fellow Ranger, a brother in arms, and the brother of the deceased. He was also a member of the 2nd Platoon, Serial #2, who was riding in the rear vehicle of the convoy while manning an inoperable MK-19. The heavy machine gun on his Humvee was ripped apart by the canyon walls while trying to squeeze through one of the narrowest necks of the canyon. With his primary weapon inoperable, all he had left to defend himself with was a 9mm pistol, from which he would fire eight rounds while Serial #2 was trapped inside the canyon. Thus, Kevin Tillman understood as well as anyone possibly could just how chaotic and confusing the situation was leading up to his brother Pat's death.

What was the right thing for leaders to do in the context of the moment on the airfield in Khowst? The answer is common sense. **To tell Kevin Tillman the truth and nothing but the truth regarding what they knew at that time about his brother's death.** What follows is one of many different ways the chain of command could have and should have communicated with him:

> *"Kevin, I know there are no words I can say to assuage your grief and sorrow at this moment; all I can offer is my most heartfelt condolences and my commitment to ensure you know everything I know regarding the circumstances of your brother's death. Right now, we don't know enough to say anything definitively; however, based on feedback from other members of the platoon and physical evidence we found at the site, we do know enough to say that Pat got hit while he was caught in the crossfire between Serial #2, Serial #1, and*

*the enemy. The only eyewitness we have is the Ranger private who was next to him, and he's pretty shaken up, but he's adamant that Pat was hit by **friendly fire**. Until we hear back from the medical examiners, we won't know for sure whether the bullets that actually killed Pat were from **friendly fire** or the enemy. Until then, I want you to know that I'll pass on everything I learn to you as quickly as I get it. I know you have a lot going through your mind right now, so if you have any questions or just want to talk to someone after you get to Bagram, please don't hesitate to call me directly. The entire battalion and all of its resources are here to support you and your family in any way we can."*

After flying back to Bagram Airfield with his brother's body on the night of the twenty-third, Kevin Tillman talked on the phone with his mother, Mary, and Pat's wife, Marie, during which he shared everything he knew about the sequence of events leading up to Pat's death.

Now consider how fundamentally different the situation might have been in the weeks, months, and years that followed if the truth had been said out loud and shared with Kevin Tillman before he left the forward operating base in Khowst. From that moment forward, every member of the Tillman family would have known what the military knew at the time: that **Pat Tillman was caught in a crossfire between friendly and enemy forces and was most likely killed by friendly fire during one of the most chaotic and confusing firefights imaginable.**

As mentioned, there are no words that can assuage the devastation and the psychological toll taken on any individual who loses a loved one for any reason. Learning that the loved one may have been killed by friendly fire certainly does nothing to lessen the devastation. Learning that you were lied to about the fact that it was friendly fire can do nothing but make it worse.

When you take a stand on a foundation of lies and half-truths, you are choosing to stand on hollow ground. The more lies and half-truths you add, the more pressure is put on the hollow foundation, until always, inevitably, the foundation collapses and everything that's built upon it crashes to the ground.

CHAPTER 9

The Aftermath and Discipline

23 April, Forward Operating Base, Khowst, 4:15 A.M.: When the three Toyota Hiluxes carrying eighteen members of the reconnaissance platoon arrived back at the Rangers base camp in Khowst, it was the first time in over thirty days that some of them had access to hot chow, hot water, and a flat bed. Sergeant Rex describes what happened next: "After we got back to the forward operating base in Khowst, my priorities were to make sure we had accountability of all our sensitive items [weapons, radios, night vision devices, etc.), and to make sure the platoon had everything they needed to refit and recover. Me and my guys were still pissed off about what happened, so once I got the platoon settled in, I headed over to the TOC to find the operations officer/S-3. I stunk pretty bad and was still wearing my kit when I walked into his office and said, 'Sir, how could you have possibly let three separate forces move on a collision course with each other without telling them where each other were at? If the first serial hadn't run up to the high ground, they probably would have fired us up because they didn't think there were any other friendlies nearby and we were all bearded up in Afghan clothes.' All he said was, 'You're not the S-3, I am, so stay in your lane,' and then he walked away to talk to someone else.

"Our chain of command and their million-dollar space-age TOCs in Khowst and Bagram had only one reason for existing: to provide command, control, and communications for the teams in the field. And the one time when we actually needed them to do their jobs, they were MIA."

24 April, 6:15 A.M.: After arriving back at the forward operating base in Khowst, the 2nd Platoon spent the rest of the day conducting maintenance on their weapons and equipment. That evening (forty-eight hours after the firefight), the platoon gathered together for a critical incident debrief. The entire platoon attended except for the platoon leader and radio operator, who were in the hospital, and Kevin Tillman, along with one Ranger from Serial #1, who flew back to Bagram with Pat Tillman's body on the night of the twenty-third. Running the critical incident debrief were the platoon sergeant and the battalion chaplain. Conspicuously missing were any members of the chain of command.

Speaking in a calm, comforting tone, the platoon sergeant opened the meeting by telling the platoon: "Anyone who wants to speak can speak freely. This is just a chance for everyone to talk about what they experienced and what they felt. We're not here to respond to what someone says, but just to listen."

Squad Leader #1 spoke first. He described hearing the initial explosion and running up to the high ground on the spur to provide covering fire. He talked about occupying the position on the military crest of the spur. His voice cracked as he described approving Corporal Tillman's request to continue moving with the Ranger private and the Afghan to a separate position behind the boulders. "They ended up where they did because they were attached to my squad." His regretful tone carried obvious implications; Squad Leader #1 was taking responsibility for what happened under his leadership.

Squad Leader #2 spoke next. He began in a quiet monotone

as he described exiting the canyon and seeing the Afghan soldier shooting directly at or over his vehicle. He told of firing six or seven rounds at the Afghan and watching him fall. "I had tunnel vision," he conceded. His tone changed to emphatic as he explained the actions of the machine gunners in the back of his Humvee. "They did what they were trained to do and fired in the same direction and location as I did." Once again, the implication was obvious: Squad Leader #2 was taking responsibility for what happened under his leadership.

Every Ranger had the opportunity to speak about what happened during the debrief. Instead of pointing fingers and blaming each other, the Rangers of 2nd Platoon accepted responsibility for what happened even though they couldn't explain how it happened or how they could have avoided it.

Good leaders take responsibility for what happens under their leadership. There's a significant difference between taking responsibility and taking blame. Taking responsibility for what happened means accepting that you are responsible, accountable, and answerable for what happened. Taking blame means you are accepting that it was your fault.

After two hours of open, frank, painful, and mostly solemn discussion, there was no longer any doubt amongst those in the room about what had happened. "After the critical incident debrief, we understood the harsh reality: Pat was accidently killed by friendly fire," one of the squad leaders explained to me.

Which begs the question: If the platoon concluded on the evening of 24 April that Pat Tillman was killed by friendly fire, why did it take the chain of command another thirty-three days to tell Pat's brother Kevin and the rest of the family what they knew?

Virtually every leader, from the battalion commander to the Ranger regimental commander to the commanding general of their higher headquarters, told the investigators the same thing when asked why they waited so long to tell the family. They told investigators they "felt obligated to wait until a thorough investigation had been completed in order to avoid telling the family something that was not true." This all-or-nothing rationalization presupposed they had no other option, but they did. They could have used common sense and told the family the whole truth and nothing but the truth, as they learned it from the first day forward. We now know the entire chain of command knew that Pat Tillman was killed by friendly fire a few hours after the completion of 2nd Platoon's critical incident debrief on 24 April.

The battalion commander told investigators he knew it was fratricide after he walked the ground outside Mana on the morning of 23 April. He also told investigators that he called his boss (the Ranger regimental commander), who was running the CFT at Bagram, and told him "it was fratricide" at approximately noon on the twenty-third. The Ranger regimental commander testified that he immediately told his boss (the special operations commanding general) who was monitoring the situation from the CFT in Iraq.

There is no record of these conversations because the investigators never told the chain of command to produce them. However, based on information learned from a senior staff officer who was working alongside the commanding general in the Iraq CFT, we can now confirm that the commanding general knew that Pat Tillman was killed by friendly fire no later than the night of 24/25 April. The senior staff officer is a highly experienced and well respected member of the special operations community who had also served time in the 2d Ranger battalion as a lieutenant.

Senior Staff Officer: "I was talking with the commander [a one-star general at the time] in his office when the phone rang. It was his boss, the CENTCOM commander." The CENTCOM commander is a four-star general who is the overall commander of all military forces operating in both Afghanistan and Iraq. "The general spent the first few minutes of the call talking about the problems he was having working with the conventional generals in Iraq. Then his tone changed from administrative to angry and urgent: **"Sir, you need to know that we killed him; it was fratricide, and we need to get in front of the message."**

I asked the senior staff officer if the commanding general said anything about contacting Kevin Tillman and the family to let them know?"

Senior Staff Officer: "No, he didn't say anything like that. I didn't think anything about the conversation until I saw the memorial service on the internet a week or so later and wondered why they were still saying he died while assaulting an enemy position. I thought it was some kind of mix-up. When I heard they hadn't told the family I felt really bad and wanted to tell them what I knew but never got the chance.[84]

It would be another thirty-three days until Kevin Tillman and the family were told what the entire chain of command knew on day three.

25–27 April: An autopsy was conducted at Dover Air Force Base, Delaware, that concluded, based on available evidence, that the cause of death was friendly fire. Of note, the report was incorrectly dated 22 July, 2004.

27 April: Kevin Tillman arrived at San Francisco International Airport and was met by family members.

28 April: The special operations commanding general reviewed and signed the manufactured Silver Star narrative and emailed it to the acting Secretary of the Army. The email consisted of four documents: a one-paragraph "award citation" that summarized Pat Tillman's courage and valor; a five-paragraph "award narrative" that offered a more detailed account of his actions; and two brief statements from soldiers who witnessed those actions. None of these documents mentioned, hinted, or implied that Corporal

Tillman was killed by friendly fire, and the two soldiers who originally wrote the witness statements said that what was submitted was not what they originally wrote.

3 May: Two thousand people, along with hundreds of members of the press, gathered at the San Jose Municipal Rose Garden for Pat Tillman's memorial service. The highest-ranking military member in attendance was a three-star general from the U.S. Army Special Operations Command. In fairness to the three-star general, he had no access to the specific details of the firefight nor any context-specific facts from the first investigation, which was compartmented within the Ranger regiment at that time. The three-star general offered his condolences to the family and reassured them that the command would do "whatever the family needed to help them get through this painful ordeal." He then sat back down and listened to a long list of speakers who were there to pay their respects to the man they so fervently loved and admired.

The speakers included: Senator John McCain; Pat's ex-teammate, NFL quarterback Jack Plummer; and a number of Pat's former coaches and mentors. The only military speaker was a Navy SEAL who had befriended Pat during his month-long odyssey at Baghdad International Airport one year earlier. Although the SEAL had no idea what actually happened the night Pat was killed, a representative from the 2nd Ranger Battalion asked him to mention that Pat had been awarded the Silver Star. He agreed and decided to use the Silver Star Award narrative to explain what actually happened in the last few minutes before Pat Tillman's death. This is what he told the audience.

Navy SEAL: "If you're a victim of an ambush, there are very few things that you can do to increase your chances for survival, one of which is to get off that ambush point as fast as you can. One of the vehicles in Pat's convoy could not get off. He made the call; he dismounted his troops, **taking the fight to the enemy, uphill,**

to seize the tactical high ground from the enemy. This gave his brothers in the downed vehicle time to move off that target. He directly saved their lives with that move. Pat sacrificed himself so that his brothers could live."

In her poignant book *Boots on the Ground by Dusk*, Mary Tillman shared her feelings about what the Navy SEAL said: "He was the first person to give us an account of Pat's death, and it provided me with a small measure of peace which had been absent since I learned of his death seven days earlier." At the end of the ceremony, as *"Taps"* played in the background, three Rangers presented Mary Tillman with a folded American flag. The Ranger that handed the flag to her was one of the Tillman brothers' closest friends in the 2nd Platoon and a member of Serial #1, who was pinned down on the spur a short distance from where Pat took his final breaths. After the ceremony, the Tillman family invited both the SEAL and the Ranger back to their home for a gathering of friends.

During the gathering, Pat's father asked the Ranger to share whatever he could about the firefight that took Pat's life. The Ranger would later tell investigators: "I was told by the battalion sergeant major before I left Afghanistan, just to pretty much keep my mouth shut up about the incident until all the pieces were put together. I lost respect for the people in charge of me. It just really pissed me off. I saw the pain in the family's faces and knew what they were going through over what happened and I wasn't allowed to tell them the truth."

The anger and resentment he felt toward his chain of command overwhelmed his common sense, and instead of returning to Fort Lewis, Washington, the next day as ordered, he went AWOL instead. He would later tell investigators that he was "fed up with all the lies."

The next day he returned to the Tillman home to talk with Kevin, who knew he was AWOL and didn't understand why. Kevin

Tillman told investigators, "I pleaded with him to tell me what he knew and what was bothering him, but he didn't tell me anything." In the context of the moment, one of the Tillman brothers' closest Ranger friends made an emotion-based choice: to "keep his mouth shut about what happened" because he feared the repercussions from the chain of command he loathed more than the repercussions from the family he had grown to love.

4 May: The first official investigation into the death of Pat Tillman was completed. The investigating officer was a captain in the 2nd Ranger Battalion, and his findings were based on interviews he conducted with individual members of the platoon during the first few hours and days after the ambush. In his summary, he wrote:

> **Findings:** *Based on interviews, sworn statements, and physical evidence, and ascertaining all the facts to the best of my ability,* **I found that there were a number contributing factors that led to the friendly fire incident that occurred on 22 April 2004.**
>
> **It is my unbiased and impartial opinion that it was these contributing factors that led to the accidental death of Pat Tillman.** *It is important to state, that based on my findings, I do not for one moment believe those involved intentionally fired at friendly forces, nor was there criminal intent involved. However, I do believe leaders at all echelons, down to the lowest Ranger, disregarded basic tactical principles and leadership responsibilities.*
>
> *Most prevalent contributing factors were* **miscommunication from the company chain of command to the platoon leader,** *human error, lack of communications, lack of positive target identification measures, inadequate preparation at the platoon level, failure to adhere to unit SOPs, and*

*the **inability of leaders to maintain situational awareness
in a combat environment.***

As referenced earlier, the Tillman family was given the results of each of the four investigations as they were completed—with the exception of the first.

The first investigation disappeared shortly after it was completed, and the only reason it reappeared was because the investigating officer inadvertently mentioned its existence to Kevin Tillman during a casual conversation after he returned to Fort Lewis, Washington.

It turned out that the "Findings" from the first investigation were rejected by the Ranger regimental commander, who then assigned his deputy, a lieutenant colonel, to conduct Investigation #2. Once the existence of the first investigation became known, the investigating officer was ordered to turn over every document he had on file, which included all of the statements he had collected as well as all drafts of his "Findings."

While comparing and contrasting the final "official version" of the "Findings" as seen above, with the final "draft version" the Investigating Officer submitted, I noticed a major discrepancy.

In his "final draft" version, the Investigating Officer wrote:

> *Most prevalent contributing factors were human error, lack of positive communications, lack of positive target identification measures, **miscommunication from all echelons from the <u>battalion down to squad level,</u> <u>and the</u> <u>inability of all leaders to maintain situational awareness in a combat environment</u>.** (Emphasis added.)*

Whoever reviewed the final draft drastically changed the wording by striking the word ***battalion*** and replacing it with ***company,*** and striking the word ***all*** before the word ***leaders.*** No other investigation identified "battalion"-level leaders as a "contributing factor" in the death of Pat Tillman.

Here's what one former A Company Ranger said when I showed him the two versions: "How ironic that the only time the battalion leadership was blamed was when the battalion did the investigation. Sad, sad, sad."

What difference do those key words make? Whoever chose to delete the phrase, *"from battalion down to squad level,"* and replace it with, *"from the company chain of command to the platoon leader,"* and to delete the word *"all"* before *"leaders,"* **chose to deny future leaders their freedom of choice to learn from it.** If future leaders can't learn from past experiences, they can't adapt to similar situations in the future. And as 99.999% of all species that have ever walked the earth prove, if we can't adapt, we perish.

The lead investigator of Investigation #3 was a brigadier general who was unaware that the first investigation had been conducted and, hence, why it had been rejected. When he found out, he asked the battalion commander to explain it to him.

Investigator: "What was his [the Ranger regimental commander's] concern?"

Battalion Commander: "It was written, to be honest, Sir, I was pissed off at this platoon leader and this platoon. They had done a number of things wrong, and he [the captain who conducted the investigation] knew how I felt about this whole situation. And he kind of—it made him agree with it. And so I think some of that may have come through in the investigation, and maybe, perhaps, because I was implicated as being wrong, he [the Ranger regimental commander] wanted to get somebody of equal rank to take care of it."

7 May, Fort Lewis, Washington: Kevin Tillman signed back in at the 2nd Ranger Battalion compound, where he hoped to begin the healing process by taking his first steps toward getting on with his life without his beloved brother.

Although the main body of the battalion were still in Afghanistan and not scheduled to return until 24 May, the battalion commander flew back during the first week of the month. When he arrived, he was told Kevin had signed back in to the battalion, and instantly realized the hollow ground he was standing on was only a few words away from collapsing beneath him.

This is how he described the situation to investigators:

Battalion Commander: "It was the end of April, first of May, basically, and I called the regimental commander and said, 'Sir, I think we need to approach Kevin Tillman and his family with this information. We're back, and I cannot separate these guys. I mean, you've got 600 Rangers. Everybody knows the story. This is going to get out. I'd like to go ahead and do it.' So we had planned to do it after Memorial Day—whatever that date is. It was that week after Memorial Day. So it was after that, and everybody had said, 'Yeah. We'll do that.' Then the Phoenix Times or whatever they are—the Sun somehow got a hold of this story. And that came up through the press corps that they were going to release this. So that's when the USSOCOM commander [a four-star general in charge of all special operations units] got in on all of this and said, 'Hey, we need to go do something soon. You're kind of the only guy that knows the story.'

"What I did do before the press released it was bring Kevin in my office and give him the same details so that he knew. He called his family and told them why I was coming and what this was about. So they all knew, but it was not the standard notification process—I mean, to have that—I mean, Kevin and Pat were in the same platoon, so it was an interesting dynamic to work through."

21 May: The 2nd Ranger Battalion returned from Afghanistan after seven weeks on the ground in Spera District, Khowst Province. Despite the questionable nature of their original mission/purpose (Operation Mountain Storm), they had learned a lot of valuable information about the people, the terrain, the enemy, and the tactics and techniques used along the way. Notably, Sergeant Rex's reconnaissance platoon spent six weeks living off the land and establishing relationships with the locals in and around the district.

I am not aware of any other force that spent that many nights in the mountains at any point in the twenty years the U.S. was in Afghanistan. There should be a file or monograph in the infantry library with a title such as "Lessons Learned Living and Operating in the Mountains of Eastern Afghanistan." Yet the military unit that rotated into Afghanistan after the Rangers did not operate in the same area, nor did they follow the same strategy. As a result, there was no continuity of purpose between the units **and no way to sustain and build on the lessons learned as well as the relationships that were established by the reconnaissance platoon with the people of Afghanistan.**

27 May: Five weeks after Pat Tillman was killed, and three days after the battalion returned to Fort Lewis, Kevin Tillman was finally told the truth about what had happened.

Kevin Tillman: "I found out everything in pretty much one fell swoop. I showed up to work and they didn't say anything to me about it until, like, eleven o'clock or something. So I did my PT with two of the people that killed Pat and then went to breakfast with the platoon leader, who eventually got fired, and I was telling him, 'Hey, you did a good job out there,' not having a clue what really went on in that first part, so I'm trying to pump the PL up. At eleven o'clock I get called in the office, and he just kind of finger-drilled it when he said: 'We think, judging by the information, there's a good chance it was fratricide.' And it was kind of danced around. It just didn't make any sense the way I was kind of told what happened. I mean, it didn't cross my mind at all. It hadn't even crossed my mind until they told me...I thought he had been killed by the enemy while running uphill...you know, storming Normandy."

Ranger from Serial #2: "A few days after we got back, we were cleaning our weapons in the team room and heard a loud crash in the barracks, followed by a lot of screaming and yelling. It was Kevin screaming in anguish after they told him. Turns out, Kevin didn't know it was fratricide; he was told it was something else. This was the first time we heard that they hadn't told Kevin and his family that Pat was killed by friendly fire. We had known since the day after, so we had time to wrap our heads around it."

29 May 2004: The commander of the U.S. Army Special Operations Command (USASOC) at Fort Bragg, N.C., announces that friendly fire "probably" killed Pat Tillman in a terse announcement, after which he refused to answer any questions.

The Discipline:

29 June, 2004: Five weeks after returning from Afghanistan, and before the third and fourth investigations were initiated, the chain of command decided it had sufficient information to discipline the following individuals:

Squad Leader #2 was reduced in rank, fined two thousand dollars, and released from the Rangers for "violating battalion standards." The rationale given by the battalion commander was that Squad Leader #2 "failed to follow the battalion standard operating procedure of positive identification of a target before firing." Squad Leader #2 left the Army a few days after his punishment.

The three machine gunners in the lead GMV were also kicked out of the Rangers for "failure to follow the battalion standard operating procedure of positive identification of a target before firing."

The platoon leader was relieved by the Ranger regimental commander, who told the entire regiment that he was kicking him out for "dereliction of duty." The platoon leader also received a verbal reprimand from the battalion commander.

The operations officer/S-3 and the company commander both received written letters of reprimand for "failing to provide adequate command and control of subordinate units." Letters of reprimand such as these are known as "an administrative slap on the wrist" because they are not placed in the individual's permanent records, so the only people who know the reprimand exists are the issuer and the receiver. However, the company commander also received a below-average Officer Evaluation Report, which, in the era of overinflated performance ratings, is an unspoken career-killer. As a result, the company commander left the Army a few months later. The operations officer/S-3 (a major in rank at the time) was given a "top block" officer evaluation report, which enabled him to get promoted and eventually become a three-star general.

The battalion commander, the Ranger regimental commander, and the special operations commanding general, who were ultimately responsible for all decisions made leading up to the firefight as well as the adjudication of the four investigations conducted afterwards, were all promoted. The 2/75 battalion commander would go on to become a two-star general. The Ranger regimental commander was promoted to one-star general. And the special operations commander (a one-star general at the time) would go on to become a four-star general.

Many of the Rangers in A Company 2/75 took issue with the discipline and the fact that it was decided upon by the individuals who they believed were most responsible for what happened. So many Rangers complained about how various members of the platoon were disciplined that the third and fourth investigations were directed to look into the process and the purpose behind the disciplinary decisions.

Investigator: "Did you tell the Tillman family there were no good words—there was no good explanation for what the soldiers did out there in the field that day?"

Battalion Commander: "I remember that question. And I remember saying, 'Ma'am, I can't explain. I can't justify to you how something like that can happen. I can only tell you that these guys were coming through an ambush, they were being shot at by three sides, and, in those circumstances, sometimes you don't think clearly. Sometimes the confusion is so great that you start to react instead of think. And I can only tell you I think that what happened is they saw movement; they reacted by shooting; and they didn't discriminate. And, in that regard, they are wrong for what they did; and we don't tolerate that.' So, Sir, I'd say the essence of your question is yes; but I don't think I would have said exactly what you said there."

Investigator: "Did you ever hear anyone say or tell other soldiers that they should share the blame for the incident or words to that effect?"

Battalion Commander: "Sir, most of your question, I think I know what you're asking me. I'm not sure what to tell you. I would not be surprised. Here's my feeling. That when the platoon came back in [to the base in Khowst], they did their hotwash on Pat Tillman's death and, oh, by the way, shooting the platoon leader and his RTO [radio operator], and it was all somebody else's fault. It was the operations officer's fault for asking them to do something and splitting them up. It was the radio's fault of not working properly. It was the enemy's fault. It was the terrain's fault. To the degree that [Squad Leader #2] couldn't look at me and say, 'I fucked up,' and I told them, 'Hey, you guys are responsible for the life of a man. And everybody here has a piece of that responsibility.' So I'm not sure if that gets at what you're saying, Sir. I would have said that. I know I did say that on a number of occasions to the NCOs."

I asked the platoon sergeant about the battalion commander's comment regarding Squad Leader #2:

Platoon Sergeant: "He was devastated by what happened. During the memorial service we did for Pat two nights after the firefight, we were all standing at attention, and I looked over at him and noticed he was crying. When I pulled him aside and asked him if he was okay, he said, 'I think I'm responsible for what happened to Pat.' He never denied what happened. He told all the investigators he had tunnel vision. He wasn't a great communicator, and like the rest of us, he had a hard time describing with words, all the senseless orders that got him into the friendly firefight. Especially when the individuals who issued the orders are the ones you're trying to explain it to."

Investigator: "Where did you say that it was platoon's fault, or under what conditions did you say that?"

Battalion Commander: "I know I said it to [redacted]. Before I left command, I did a NCOPD [non-commissioned officer professional development]. The battalion sergeant major and I brought together all the NCOs in the battalion, and we took them through this event so that we could all understand what happened; and I don't recall saying it was the platoon's fault for sure there, but I think I would have. I felt very strongly that you've got to get over these things, and [redacted]. But at some point, you have to look yourself in the mirror and know that you made mistakes that you could have prevented. And I didn't think they did that right away."

I asked the platoon sergeant about the NCOPD:

Platoon Sergeant: "The battalion commander and battalion sergeant major did a NCOPD on the firefight, and every NCO in the battalion was told to attend, except for the NCOs in 2nd Platoon. The battalion sergeant major ordered us not to attend. None of us had any doubt about why they didn't want us there. It was because they wanted to blame the whole thing on us. After that NCOPD, it seemed like no one wanted to talk to us. We were basically shunned. We were outcasts. I didn't care about myself, but I knew how bad the platoon was hurting from this thing. They revered Pat, and every one of them were devastated by his death. There was no healing. Most of us left voluntarily; we never got closure."

Investigator: "What was the platoon leader punished for?"

Battalion Commander: "My problem with the platoon and the platoon leader was that I had taken, I thought, a lot of time with my lieutenants to teach them how to be good leaders. And I just had the ass that this guy, after all of the times I'd trained him, I thought he was irresponsible in the way he task organized and led that platoon. And we had gone through a number of LPDs [leader professional development] with platoon sergeants and platoon leaders in Afghanistan about everything from why we don't drive in the daylight, to troop-leading procedures, to truncated troop-leading procedures, and really gone through the nuts and bolts of how I wanted my leaders to do business out there. We had just had Sergeant Blessing killed out there for reasons I thought were attributable to leadership. And, in my mind, the reason they performed so poorly in contact was because they didn't do, and if I personalize it, they didn't do what I told them to do. They didn't do what we trained to do. And, Sir, I just thought that was wrong. So I let him go…And really, as you go down from me, on down the line, you can assign more and more degrees of culpability on how we could have prevented this from happening."

As you'll see in the comments from individual Rangers below, there were some differences in opinion with regards to the disciplining of the Rangers in the lead Humvee; however, **there was unanimous agreement that the platoon leader was scapegoated in order to protect the chain of command above him:**

"Everybody thinks the platoon leader got the shaft. They shit-canned the platoon leader for splitting the platoon, even though he didn't want to split it at all. But because he was responsible for the platoon overall, he was booted out of the Rangers."

"It was a complete insult and a total scapegoat to send this guy on his way when he initially didn't agree with the command that was given to him."

"He did the best he could in the situation; and then he gets let go because of why? They needed someone to place the blame on."

"If the Army has to decide whether to punish a lieutenant colonel at headquarters or a lieutenant in the field, you better believe the lieutenant's going to take the hit every time. Shit rolls downhill."

What follows are comments made by individual Rangers when asked what they thought about the way Squad Leader #2 and his men were disciplined.

Sergeant Major (ret.) Ted Kennedy: "A common trait of inexperienced leaders and soldiers is wanting to shoot and make noise but not thinking about what's actually going on around them. This gun-mounted Humvee had more firepower than some light infantry platoons, yet the leader chose to shoot and not control. Leaders are not fighters unless they are put in a position to defend 'theirs.' I fought the culture of 'shoot first, ask questions later' as a first sergeant after I learned the lesson in Panama as a twenty-one-year-old squad leader with A Company. I often argued that one bullet into one head was more effective and way safer than a thousand rounds that missed. Unfortunately, the battalion commander did not see it that way...he was a 'volume of fire' sort of national training center fanatic, and that's the way he setup all of our training events. His philosophy rubbed off on other leaders in the battalion who liked to hear gunfire and equated that to effectiveness. 'We fight as we train!!'"

Former 2/75 Command Sergeant Major: "It's hard to understand how they could punish the Rangers in the lead Humvee for shooting at the spur without positively identifying their targets, when the four Rangers in the second vehicle did the same thing. Didn't that prove how difficult it was to positively ID friendlies on the spur?"

Corporal Tillman's Organic Squad Leader (Riding with Serial #2, in the Jinga): "To the best of my knowledge, [Squad

Leader #2] saw the Afghan soldier firing with his weapon pointed to the south, and he engaged the Afghan soldier. The truck engaged everybody else in the vicinity of the Afghan soldier. If I was in the same situation and I saw somebody who was shooting at me and my men, I would engage him also. And unfortunately, in this situation, the worse situation happened and in the worse terrain. You had a force maneuvering to save another force, and neither one of them were aware where the other force was they were engaging. Tillman, who was one of my team leaders, got maneuvered into an area that was being engaged."

Platoon Sergeant: "They were kicked out of the Rangers for 'failure to follow the battalion SOP of positive identification of a target before firing.' But for a standard operating procedure to become standard, it must be trained and reinforced, and we never conducted any Humvee maneuver live-fires with friendly and enemy targets intermixed. We never trained these guys how to discriminate between friendly and enemy targets while moving and shooting from a Humvee."

Ranger Private: "It's still my belief that the guys in the GMVs aren't to blame for it, Sir. I don't think at all what happened to them was right. I don't feel that they did anything wrong, and I was right there. You know, it should have been me dying instead of Pat, and I can still say that those guys in that GMV were some of the best Rangers I know, and I don't hold them responsible for anything, and I know Pat wouldn't either."

Note: Almost twenty years later, the Ranger private whose life was saved by Corporal Tillman was still in the Army and still leading elite airborne soldiers as a Battalion command sergeant major. Command sergeant major is the highest rank achievable for an enlisted soldier and one of the most important and impactful leadership positions in the military. As **Command Sergeant Major Bryan O'Neal** explained to me, "Pat gave his life to save mine, and I think about him and the life lessons he taught me every single day. I want Pat to be proud of what I've accomplished, so I never forget how lucky I am to be alive nor my responsibility to pass on what I learned."

When I asked Command Sergeant Major O'Neal if he could give me some examples of the lessons he learned from Pat, here's what he told me: "First is to 'always take care of the dudes and dudettes,' which was Pat's way of saying 'treat other people like you want to be treated.' One day Pat told me to string some chem-lights together for a training exercise we were doing the next day, and I totally forgot to do it. The next day, when he came to work and found out I didn't do it, he looked at me and said, 'You really disappointed me.' I was crushed. I wanted to say, 'Please make me do some push-ups or something,' because knowing he was disappointed in me was the worst thing that could have happened. Pat always treated everyone with total respect; he believed that people will do more when they want to instead of have to. The other thing he always emphasized was the importance of staying physically fit. He'd say, 'You can't be a competent infantryman unless you are in top physical condition, if you can't carry your weapon and ruck to the target, you are irrelevant; if you can't carry your buddy off the target, you are irrelevant; if you can't lead from the front, you are irrelevant.'

I also tell my soldiers what we learned from the firefight. Expect the unexpected, and when the shooting starts, get down."

Leading teams in the Army over the last 20 years has enabled Sergeant Major O'Neal to pass on these lessons and positively influence the lives of countless soldiers. Pat Tillman saved his life that night. Today, Pat's legacy lives on in all of us.

Conclusion

I trust your opinion, so please let me know what you think happened and how we can keep it from happening again.

To Mary Tillman: The army leadership failed you, but Pat did not. Pat gave his life to save another. He earned his silver star, and you can be proud of his actions and his legacy, which will live on for a long time. From the beginning, your motherly instincts (aka common sense) alerted you that the story they were telling you didn't make sense. You ignored the naysayers and bureaucratic barriers, and you refused to accept that "there's nothing else to learn." Your resilience and dedication toward finding the truth are an inspiration to every citizen, every soldier, and every parent. When we first talked, you told me, "If the situation was reversed and it was me that was killed, Pat would never give up." As the son of an amazing mom myself, I believe that Pat would be very proud of you and very grateful for what you've done. I started this endeavor to find out **what happened, and what the rest of us can learn from this tragic situation** to further our own knowledge and add to the safety of fellow first responders in the future. What follows are my thoughts and recommendations:

THOUGHTS

1) What happened? As the four investigations, hundreds of pages of sworn statements, and my own interviews with the men of the 2nd Ranger Battalion reveal, **Pat Tillman's death was the result of a tragic friendly fire accident.**

Yet the friendly fire accident didn't just happen due to the fog of war, battlefield friction, or enemy acumen. Instead, the friendly fire accident emerged over time from the sum total of senseless choices made by the chain of command and the toxic leadership climate those choices created:

> *Don't leave your inoperable Humvee in a secure base (BCP#5) where it can't cause you anymore problems; drag it with you through enemy-occupied mountains where it can. Permission to blow up the vehicle is denied, but we don't care what happens to it, just tow it along with you until you can't. Split your platoon, and here's how you need to do it. Get boots on the ground in Mana before nightfall. Do what you're told to do. We're too busy to talk right now; we'll get back to you when we can.*

The conflicts that cause the most stress are those between what the strategy, plan, or disconnected chain of command is telling us to do and what the reality of the situation going on around us reveals we should do. The platoon was against every one of the above decisions because none of them made sense based on the reality of the situation going on around them. Despite their common sense objections, the platoon was ordered to follow them anyway.

Our purpose paves our choice of paths. **For our choice of paths to make sense, our purpose must make sense.** When our purpose doesn't make sense—*split the platoon to get boots on the ground* *before nightfall*—the paths we choose to accomplish it can't and won't ever make sense. Blaming the platoon for what happened is like blaming someone for falling down after their legs were chained together. Don't blame the victim. Blame the chain.

The senseless orders, directives, and lack of support by the "chain" made the Rangers of 2nd Platoon do things they never would have considered if they had been given the **freedom to adapt to the situations going on around them.**

2) What can the rest of us learn from this tragic situation to further our own knowledge and add to the safety of fellow first responders as similar situations unfold in the future?

That the common enemy of friendly fire and toxic leadership is common sense. Common Sense Leadership Matters. Toxic Leadership Destroys. We must teach leaders how to use common sense to take care of their people by making good decisions and solving complex problems that set the conditions for their people to succeed.

RECOMMENDATIONS

Before getting to the recommendations it's important to acknowledge and remind ourselves of the high cost of freedom and the price that's paid by the first responders who volunteer to serve and protect our freedoms. Freedom-loving societies only survive in a world of violence, tyranny, and sedition when there are people willing to fight and die for freedom on others' behalf.

> *As we honor their memory today, let us pledge that their lives, their sacrifices, and their valor shall be remembered, not just with words, rather with actions in support of what must have been their wish: **that no other generation of young men will***

ever have to share their experiences and repeat their sacri-
fice." Ronald Reagan, Memorial Day, 1982

Recommendation #1: The Army owes the Tillman family an apology. The Department of Defense should officially apologize to the family in a written correspondence that covers all the failures that occurred and what they are doing to ensure those failures never happen again.

Recommendation #1a: The Army owes the Rangers of 2nd Platoon an apology. The 2nd Platoon was unfairly scapegoated by their chain of command and they knew it, yet their inability to understand and explain what actually happened and why left them with feelings of frustration and guilt, the twin pillars of PTSD. The Army should re-examine the disciplinary actions handed out by the chain of command and use the new evidence contained in this book to correct the record.

Recommendation #2: Correct the record on the false Silver Star narrative. I am petitioning the Department of Defense to correct the record and replace the fraudulent Silver Star narrative with an accurate one that reflects both the reality of the situation on the spur and the reality of Corporal Tillman's combat leadership and common sense initiative. I will proudly volunteer to write the new narrative based on what actually happened so that everyone who reads it understands how deserving he is and so his family has something they can believe and take pride in.

Recommendation #3: Make the Logic of Why and Saying it Out Loud Mandatory: When issuing orders, commands, or directives in combat, leaders must say out loud the logic of why their orders do or don't make sense. Saying it out loud makes our thoughts physical, which enables other senses—our own and those of the people around us—to pressure test what we're thinking and see whether it makes sense or is senseless. The more senses we involve, the more sense we can make.

*Technology hasn't made saying it out loud
obsolete; it's made it absolute.*

Your neocortex (thinking brain) is the only part of our brains that can create and process language. It's also the only part of our brain that can pay attention to what's going on around us and make sense of it. The biology of our brains explains it. To pay attention to what's going on around us, our neocortex needs a purpose. The logic of why provides our neocortex with the logic of purpose. It explains why we do what we do and choose what we choose. When striving to understand the choices we make, it's not the choice itself (yes/no, buy/sell, etc.), but the "logic of why" that we used to make the choice that explains, communicates, and validates whether the choice makes sense or is senseless. The logic of why replaces orders and commands, and, when said out loud, provides checks and balances against senseless chains of command. (To learn more information about the logic of why, see page 132 in this book, or Book II, *The Common Sense Way*, Chapter 5.)

Recommendation #4: Believe in and trust your senses: We aren't helpless or dependent on our chains of command for making sense of what's going on around us. Our secret weapon for understanding the world around us is our senses. Believe in and trust your senses. To engage them, all we have to do is take a couple long, full, deep breaths and pay attention to what's going on around us. Stay calm think. Like a cat searching for a mouse, look, listen, smell, touch, feel, then say out loud what you perceive and share it with others to see whether or not it makes sense.

Recommendation #5: Change the way we think about toxic leaders and the toxic leadership climates they create. We have always known about toxic leaders and toxic leadership climates, but we never connected them to the catastrophes they create. Until I researched this incident, I never thought of toxic leadership climates in the context of mission failure or casualties. This disconnect was most likely due to the delayed reaction between the toxic choices leaders make and the harm those choices inflict on their people and organizations. One of the unexpected outcomes of examining a toxic leadership climate twenty years later is that the scope of devastation it wreaks becomes more apparent. At least two of the Rangers I interviewed were still battling suicidal thoughts. Toxic leadership destroys. It's no longer theoretical. As mentioned in Chapter 4, the military must make eradicating toxic leadership from its ranks a tangible priority by changing its current culture of toxic tolerance to one marked by toxic intolerance.

Recommendation #6: Initiate a PTSD Study of Groups: How many other military units have such high PTSD rates caused by a shared mission, battle, or tour of duty? What can these clusters tell us about PTSD? Does the way that an organization handles a traumatic event affect the potential for individuals to get PTSD? Is there a correlation between PTSD and the leadership climate in those units? Can PTSD be treated by group therapy, where the individuals get together and talk it out by learning about what really happened?

My experience interviewing many of these men made me realize how important it is to talk with, listen to, and share knowledge with them. What follows is a text message sent to me by one of the Rangers a few moments after we finished a two-hour conversation:

"Bottom line. Talking with you, even about the worst 10 minutes of my life, albeit painful...helped. I owe you, Sir. The darkness in my head is real. Thanks for hitting the white light on the Surefire to allow me to get some perspective."

- 11–23% of veterans experience PTSD within any given year.
- Statistically, 17.2 veterans die by suicide each day, and are 1.5 times more likely to die by suicide than civilians.
- 3 in 10 (30%) of first responders suffer from PTSD.

As part of their PTSD program, the military also needs to initiate regularly scheduled training on stress inoculation and how to recognize and respond to cortical inhibition. (See Chapter 7)

Recommendation #7: Establish a Friendly Fire University. We need to stop sweeping friendly fire under the rug and start talking about it. Specifically how to avoid and prevent it. The operating focus of the Friendly Fire University should include the following:

- Train and teach to create awareness of friendly fire.
- Investigate current and past incidents.
- Engage in research & development focused on preventing friendly fire.

The organization should be run by veteran first responders with military, police, or firefighting backgrounds. It must be independent from the Department of Defense, preferably a non-government, non-profit organization that is funded by grants and donations. The goal is to create a credible and trusted place where civilians and military can contact, report, and get information about friendly fire.

Recommendation #8: Establish the Common-Sense Leader's Oath: Upon taking a leadership position, all leaders must say out loud the Common-Sense Leader's Oath: I solemnly swear to use common sense to take care of the people I have the privilege to lead by making good decisions and solving complex problems that set the conditions for my people to succeed.

For live updates on the status of these
recommendations go to peteblaber.com

ANNEX A

Military Unit Size and Leadership Rank

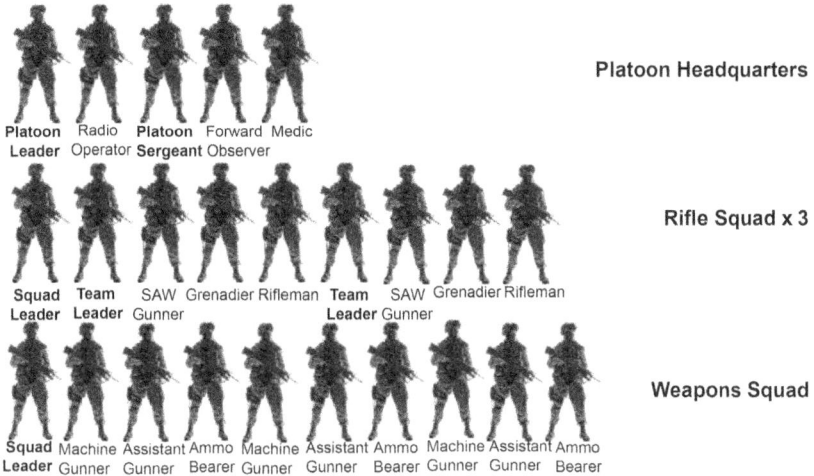

Platoon Headquarters

Platoon Leader · Radio Operator · Platoon Sergeant · Forward Observer · Medic

Rifle Squad x 3

Squad Leader · Team Leader · SAW Gunner · Grenadier · Rifleman · Team Leader · SAW Gunner · Grenadier · Rifleman

Weapons Squad

Squad Leader · Machine Gunner · Assistant Gunner · Ammo Bearer · Machine Gunner · Assistant Gunner · Ammo Bearer · Machine Gunner · Assistant Gunner · Ammo Bearer

Military Unit Size and Rank of Leadership

Unit	Number of Soldiers	Rank of Leaders
Team	3–5	Corporal or Sergeant (E-4 & E-5)
Squad	5–10	Staff Sergeant (E-6)
Platoon	30–40	Lieutenant & Sergeant First Class (E-7)
Company	90–120	Captain & First Sergeant (E-8)
Battalion	450–650	Lieutenant Colonel & Command Sergeant Major (E-9)
Regiment/Brigade	1,350–1,950	Colonel & Command Sergeant Major
Division	4,000–6,000	Major General (2 Stars) & Command Sergeant Major

ANNEX B

Ranger Weapons Information

M249 SAW (Squad Automatic Weapon)
Manufacturer: FN Herstel, Belgium
Used by: USN, US Army, USMC, USAF
Weight: 18 lbs
Length: 40.74 in
Barrel length: 18 in
Caliber: 5.56x45 mm
Action: gas-operated, open bolt
Max Rate of Fire: 800 rpm
Muzzle velocity: 3,000 ft/s
Max Range: 3,600 m

M240B Machine-gun
Manufacturer: FN Herstel, Belgium
Used by: USN, US Army, USMC, USAF
Weight; (M240B) 27.1 lbs; (M240G) 25.6 lbs;
Length: 49.7 in
Barrel length: 24.8 in
Caliber: 7.62×51mm
Action: Gas-operated, open bolt
Max Rate of Fire: 950 rpm
Muzzle velocity: 2,800 ft/s
Max Range: 3,725 m

M2 .50 Caliber machine gun
Manufacturer: Various
Used by: USN, US Army, USMC, USAF
Weight: 84 lbs
Length: 65 in
Barrel length: 45 in
Caliber: .50 BMG
Action: Short recoil operated
Max Rate of Fire: 850 rpm
Muzzle velocity: 2,900 ft/s
Max Range: 6,800

MK19 40mm machine gun
Manufacturer: General Dynamics Armaments
Service: USAF, USMC, US Army, USN
Caliber: 40mm
Weight: 77.6 pounds without mount or tripod
Length: 43.1 inches
Max Range: 2,212 m

M4 with ACOG gun-sight
Manufacturer: FN Herstel, Colt
Services: USN, US Army, USMC, USAF
Weight: 6.36 lbs
Length: 33 in (stock extended)
Barrel length: 14.5 in
Caliber: 5.56x45 mm
Action: gas-operated, direct impingement
Max Rate of Fire: 950 rpm
Muzzle velocity: 2,900 ft/s
Max Range: 600 m

The AT4 is a Swedish 84 mm unguided, man-portable, disposable, shoulder-fired recoilless anti-tank weapon manufactured by Saab Bofors Dynamics. The AT4 is not a rocket launcher strictly speaking, because the explosive warhead is not propelled by a rocket motor. Rather, it is a smooth-bore recoilless gun.

NOTES

1   *The American Heritage Dictionary* defines a case study as follows: 1) A detailed, intensive study of a unit, such as a corporation or a corporate division, that stresses factors contributing to its success or failure. 2) A detailed analysis of a person or group, especially as a model of medical, psychiatric, psychological, or social phenomena. 3) An exemplary or cautionary model; an instructive example.

2   https://pattillmanfoundation.org/

3   Hackworth, David H. *About Face: The Odyssey of an American Warrior,* 1989, Pan Macmillan

4   The 2nd Ranger Battalion dates back to WW II, when they led the assault on the beaches of Normandy on D-Day. In the movie *Saving Private Ryan,* actor Tom Hanks played the part of a common-sense commander in the 2nd Ranger Battalion.

5   **A note to the reader on Chapter One:** This chapter was originally published in my second book, *The Common Sense Way,* in 2021. It was written as the companion chapter to the second section of this book on what really happened to Pat Tillman and his platoon in Afghanistan. However, due to COVID-19 and the worldwide shutdown that followed, I wasn't able to travel and meet with a select group of former Ranger non-commissioned officers (NCOs) who had agreed to contribute to and fact-check the second section. Since the book had already been formatted, I had to publish this chapter alone as part of my second book. The story and the lessons learned from this chapter concerning "How Common Sense Leaders Create Healthy Leadership Climates" provide foundational knowledge for understanding and contrasting with the second section of this book, "What Really Happened to Pat Tillman and His Platoon in Afghanistan." So instead of asking or urging you to go buy my second book to read this chapter, I simply included a new and updated version in this book for free (these are the types of things you can do when you self-publish). If you have already read the previous version of this chapter, I first want to say thank you, and I also want to encourage you to re-read the updated version or to simply read the revised lessons learned in Chapter 2, titled "What a Healthy Leadership Climate Looks Like," as a refresher (pages 30–36).

6   Hackworth, David H. *About Face: The Odyssey of an American Warrior,* 1989, Pan Macmillan

7   "Drone" in this context means to behave in a dull, drowsy, or indifferent manner.

8   A position established when a patrol halts for an extended period of time to rest or prepare for a follow-on mission.

9   Sergeant X is not his real name. As you will learn in the next section, he is a part of both stories. He served in the military with distinction for thirty years and retired as a sergeant major. After he retired from the military, he continued serving the country in a different operational capacity. He asked me to use his pseudonym (Sergeant X) throughout the book.

10   A handrail is a linear feature you can move parallel to, such as a stream, a trail, a ridge, or a floodplain.

11   *The Common Sense Way,* 2021. The logic of why is the logic of purpose; it informs why we do what we do and choose what we choose.

12   *The Common Sense Way,* 2021. A sensible choice is a choice that coheres with our common purpose, and adapts to the situation going on around us. A coherent-adaptive choice is a sensible choice.

13   Ulmer, Walter, LTG. "Leaders, Managers, and Command Climate." July 1986, Armed Forces Journal International.

14   IBID

15   The Battle of LZ X-Ray (14-19 November 1965) was the first major battle between the U.S. Army and the People's Army of Vietnam (PAVN), as part of the Pleiku Campaign conducted early in the Vietnam War, in the central highlands of Vietnam. The battle is notable for being the first large-scale helicopter air assault, and also the first use of B-52 bombers in a tactical support role. LZ X-Ray set the blueprint for the Vietnam War, with the Americans relying on air mobility, artillery fire, and close air support, while the PAVN neutralized that firepower by quickly engaging American forces at very close range.

16   While researching this book, I discovered we were both born and raised in the same town: Oak Park, Illinois.

17   **Declassification is the process applied to documents that used to be** classified as secret ceasing to be so-restricted, often under the principle of freedom of information.

18 Four investigations of Pat Tillman's death were conducted by the U.S. government:

- **The first investigation:** Conducted by a Captain in the 2nd Ranger Battalion, was completed in two weeks. It concluded that: Friendly fire was the most likely cause of Tillman's death. No "criminal intent involved" in firing, but there was "gross negligence."

- **The second investigation:** Conducted by a Lieutenant Colonel from the 75th Ranger Regiment HQ, was completed in eight days. It concluded that Pat Tillman's death was "the result of fratricide during an extremely chaotic enemy ambush." Contributing factors: insufficient command and control measures, failure to execute fire control, and failure to positively identify targets as friend or foe. The chief investigator appeared to reserve his harshest judgments for the lower-ranking Rangers who did the shooting rather than the higher-ranking officers who oversaw the mission.

- **The third investigation:** Conducted by a Brigadier General from US Army Special Operations Command, was completed in two months. It concluded Pat Tillman died as a result of friendly fire. Tillman was likely struck by American 5.56 mm or 7.62 mm rounds. Failure to immediately tell the family of the suspected fratricide based on a desire to avoid giving an inaccurate picture prior to completing an investigation.

- **The fourth investigation:** Conducted by the Inspector General of the U.S. Department of Defense, Completed in 20 months (2 June 2005–26 March 2007). It concluded that each of the prior three investigations were deficient, thus "contributed to inaccuracies, misunderstandings, and perceptions of concealment." Recommended nine officers, including four generals, be disciplined for missteps in the wake of the friendly fire incident, however this recommendation was rejected and only one General Officer were disciplined. An accompanying probe by the Army Criminal Investigation Command found no evidence of negligent homicide or aggravated assault on the part of the shooters.

19 I was not the addressee on the report. This is a recreation of the flash report that I read. The order of words may have been slightly different.

20 EPW = enemy prisoner of war.

21 The U.S. Army recently announced its intention to replace the term "friendly fire" with the term "blue on blue" incident.

22 As of April 2003.

23 Standoff distance is a security term that refers to measures to prevent unscreened and potentially threatening people and vehicles from approaching within a certain distance of a building, car, or other location.

24 Article 15 procedures in the U.S. military are a form of non-judicial discipline conducted by unit commanders. These are the most common type of disciplinary proceedings in the armed forces, and they do not result in a criminal record.

25 Company grade officers include the ranks of lieutenant and captain.

26 Blaber, P. E., *The Common Sense Way*, 2021, Pete Blaber.

27 https://www.statista.com/statistics/263798/american-soldiers-killed-in-iraq/

28 29 May 2004: Army acknowledges friendly fire "probably" killed Pat Tillman in an announcement issued by the Commanding General of Army Special Operations Command, at Fort Bragg, N.C.

29 Coll, Steve, "The Errors That Killed Pat Tillman," 6 December 2004, Washington Post

30 According to U.S. Army Regulation 10–44, the mission of the Army War College is to prepare military, civilian, and international leaders for the responsibilities of strategic leadership; educate current and future leaders on the development and employment of land power in a joint, multinational, and interagency environment; conduct research and publish on national security and military strategy; and engage in activities in support of the Army's strategic communication efforts."

31 Reed, George, and Bullis, Craig, "Toxic Leadership." A Report to the Secretary of the Army, February 2003, U.S. Army War College

32 Coined by Marcia Whicker in 1996

33 Reed, George

34 Pelletier, Kathie L., "Leader Toxicity: An Empirical Investigation of Toxic Behavior and Rhetoric," *Leadership* 10, no. 4 (November 2010): 377–378.

35 http://usacac.army.mil/cac2/Repository/CASAL_ParticipantFeedback_2011_1.pdf

36 Elle, Colonel Stephen A., "Breaking the Toxic Leadership Paradigm in the U.S. Army," by 2012.

37 Sinek, Simon, page 66.

38 Reed, George

39 Elle, Colonel Stephen A., "Breaking the Toxic Leadership Paradigm in the U.S. Army," 2012.

40 The Boeing CH-47 Chinook is an American twin-engine, tandem-rotor heavy-lift helicopter. Its primary roles are troop movement, vehicle and artillery movement, and battlefield resupply. The MH-47 is the special operations version of the CH-47. For opsec purposes, this book will use "CH-47" to describe all of them.

41 https://www.frontlineclub.com/the_end_of_cop_spera/ The Frontline Club is a gathering place for journalists, photographers and other likeminded people interested in international affairs we Champion Independent Journalism & Freedom of Speech, we rally the protection of Press Freedom and fight for the safety of Freelancers in doing their important work around the globe.

42 Joe Biden's statements during the Vice Presidential Debate with Paul Ryan, 11 October 2012.

43 WASHINGTON—Secretary of Defense Robert Gates has given the new U.S. commander in Afghanistan 60 days to conduct another review of the American strategy there, the fifth since President Barack Obama took office less than five months ago. The Defense Department announced Monday that Gates has ordered the new U.S. military commander in Afghanistan, Army Gen. Stanley McChrystal, and his deputy, Lt. Gen. David Rodriguez, to submit a review of the U.S. strategy within 60 days of their arrival in Afghanistan.

44 Krakauer, Jon. *Where Men Win Glory: The Odyssey of Pat Tillman* (p. 312). Knopf Doubleday Publishing Group. Kindle Edition

45 IBID.

46 "Disconnected hierarchies" refer to chains of command that aren't directly connected to the situation/environment. The CFT in Khowst was physically disconnected from the situation going on in Magara, as well the Rangers for whom they were making life-or-death decisions.

47 Jinga trucks are five-ton diesel rigs used to transport almost all commerce around Afghanistan (e.g., people, lumber, and supplies of all types).

48 This explanation also includes what he told the Tillman family when they asked him, "Why they didn't just blow it up?"

49 Cobra II.

50 Mike Kershaw was also in the 2nd Ranger Battalion with me as a lieutenant in the late 80s. Another byproduct of the same common-sense leaders and the healthy leadership climate they created.

51 Saddam's "Men of Sacrifice."

52 The Sabot is a non-explosive tank round that consists of a narrow
    metal rod made of depleted uranium that penetrates armor and then
    disintegrates into a spray of metal fragments.

53 The platoon leader would later explain to investigators that the
    battalion SOP for clearing villages was to always meet with the village
    elders the day before to coordinate, and never clear a village at night.

54 http://www.military.com/equipment/high-mobility-multipurpose
    -wheeled-vehicle-hmmwv

55 A fragmentary order (FRAGO) informs units that one or more
    elements of the base order have changed.

56 The Mk 19 (pronounced as "mark nineteen") is a belt-fed, fully
    automatic weapon that fires 40 mm grenades at a cyclic rate of 325
    to 375 rounds per minute. The Mk 19 can launch its grenades at a
    maximum distance of 2,212 meters, though its effective range to a
    point target is about 1,500 meters.

57 The American Institute of Stress, https://www.stress.org/what-is-stress

58 https://biology.stanford.edu/people/robert-sapolsky Dr. Robert
    Sapolsky, a professor of biological sciences and neurology at
    Stanford University.

59 From the Department of Biological and Clinical Psychology at
    Friedrich Schiller University in Germany.

60 Yehuda R., Southwick SM, Nussbaum G., Wahby V., Giller E.L. Jr,
    Mason J.W. 1990. "Low Urinary Cortisol Excretion in Patients with
    Posttraumatic Stress Disorder," *J Nerv Ment Dis* **178**: 366–369.

61 https://www.heartmath.com/blog/health-and-wellness/what-you
    -need-to-know-about-stress/ "What You Need to Know About Stress,"
    Posted on March 25, 2014 by HeartMath

62 https://www.youtube.com/watch?v=AYFZAYenR20 Dr. Sapolsky
    created a documentary called *Stress: Portrait of a Killer,* in it he shares
    that the following factors increase our vulnerability to stress:

63 The American Institute of Stress, https://www.stress.org/what-is-stress

64 https://royalsocietypublishing.org/doi/10.1098/rspb.2010.2325
    "Towards a scientific concept of free will as a biological trait:
    spontaneous actions and decision-making in invertebrates,"
    Björn Brembs, Published:15 December 2010

65 IBID, Sapolsky

66 "Retirement and Socioeconomic Differences in Diurnal Cortisol: Longitudinal Evidence From a Cohort of British Civil Servants," Tarani Chandola, Patrick Rouxel, Michael G. Marmot, Meena Kumari. *The Journals of Gerontology: Series B*, Volume 73, Issue 3, March 2018, Pages 447–456, https://doi.org/10.1093/geronb/gbx058 Published: 05 May 2017

67 "Posttraumatic Stress Disorder in Adults: Epidemiology, Pathophysiology, Clinical Manifestations, Course, Assessment, and Diagnosis," https://www.uptodate.com/contents/posttraumatic -stress-disorder-in-adults-epidemiology-pathophysiology-clinical -manifestations-course-assessment-and-diagnosis

68 Trevena, J. Miller, J. (2010), "Brain preparation before a voluntary action: Evidence against unconscious movement initiation." Consciousness and Cognition. 19 (1): 447–56.

69 A slot canyon is a very narrow gorge with steep, high walls, often made from soft rock such as basalt or sandstone. Many boast waterfalls at their tail ends. Deeper than they are wide, these canyons can be treacherous for those who enter—they are prone to flash floods and offer few ways in or out.

70 As the driver was unvetted, it didn't make sense to explain the specific route to him while they were in Magarah, where he could share it with others.

71 In an actual call, they would use grid coordinates to report their location, not place names.

72 "Break" is a term used by tactical forces to break up lengthy messages of over eight to ten seconds. Keeping transmissions short makes it harder for enemy intelligence to direction find or jam the transmission. However, its most important purpose is to provide an opportunity for other stations to transmit and "break" into the conversation to report high priority messages such as enemy activity, requests for fire support, or medical evacuation.

73 Goff, Stan, "How Pat Tillman Died," August 10, 2007, https://www .counterpunch.org/2007/08/10/how-pat-tillman-died/

74 The basic load for an M203 gunner is thirty-six rounds. The maximum effective range of the M203 grenade launcher is one hundred and fifty meters for point targets and four hundred meters for area targets.

75 CH6 stands for "charge six" and refers to the amount of propellant required to fire the mortar at different ranges.

76 Blast overpressure (BOP), also known as high energy impulse noise, is a damaging outcome of explosive detonations and firing of weapons. Exposure to BOP shock waves alone results in injury predominantly to the hollow organ systems such as auditory, respiratory, and gastrointestinal systems.

77 https://www.nga.mil/resources/FalconView.html FalconView is a Windows mapping system that displays various types of maps and geographically referenced overlays. It has been used extensively by US military reconnaissance personnel since the mid-90s.

78 To see video of entire drive through slot canyon go to website: www.firefight.commonsenseway.com

79 Slepecky N. Overview of mechanical damage to the inner ear: noise as a tool to probe cochlear function. Hear Res. 1986.

80 https://johntreed.com/

81 IBID

82 The hand and arm signal for "ceasefire" is to raise the hand in front of the forehead, palm facing out, and swing the hand and forearm up and down several times in front of the face. This hand and arm signal has been the same since I joined the military in the 1980s. In all that time, I have never heard of a real-world example where it was used and it worked. I do not believe this hand and arm signal makes sense for anything but close-range signaling. I recommend that the military update their ceasefire hand and arm signal. https://armypubs.army .mil/epubs/DR_pubs/DR_a/pdf/web/ARN2747_TC%203-21x60%20 FINAL%20WEB.pdf

83 "Eagles" is the code word for friendly combatants; "crows" is used for enemy combatants.

84 He was able to talk with Mary Tillman and tell her what he knew in December of 2022.